KT-133-054

A FIREFLY BOOK

Published by Firefly Books Ltd.

Published originally under the title *Destination Mars*,
ISBN 2-263-03106-5

Publisher Cataloging-in-Publication Data (U.S.)
Dupas, Alain.
Destination Mars / Alain Dupas._1st ed.
[208] p. : col. ill., photos. ; cm.
Includes index.
Summary: Illustrated history of the search for life on Mars combined with a fictional journey of the first humans to travel to Mars in the future.
ISBN 1-55297-934-21. Mars (Planet)—Exploration. I. Title.
523.46 21 QB641.D87 2004

National Library of Canada Cataloguing in Publication
Dupas, Alain
Destination Mars / Alain Dupas.
Includes index.
ISBN 1-55297-934-2
1. Mars (Planet) 2. Mars (Planet)--Exploration.
I. Title.
QB641.D87 200 523.43 C2004-902059-5

Published in the United States in 2004 by
Firefly Books (U.S.) Inc.
P.O. Box 1338, Ellicott Station
Buffalo, New York 14205

Published in Canada in 2004 by
Firefly Books Ltd.
66 Leek Crescent
Richmond Hill, Ontario L4B 1H1

Printed in Italy

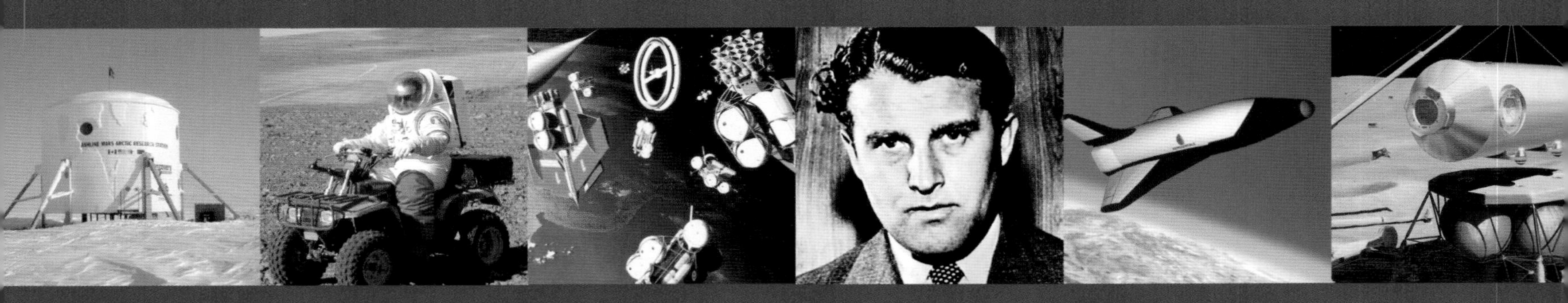

DESTINATION MARS

Alain Dupas

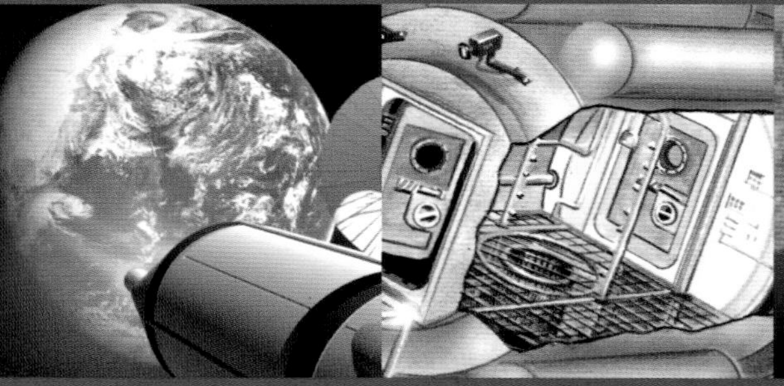
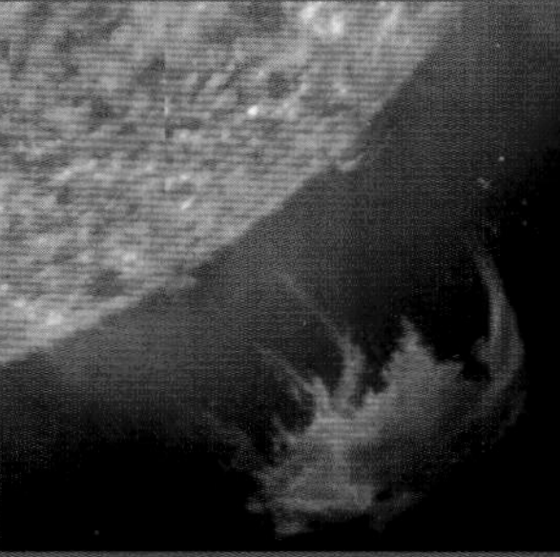
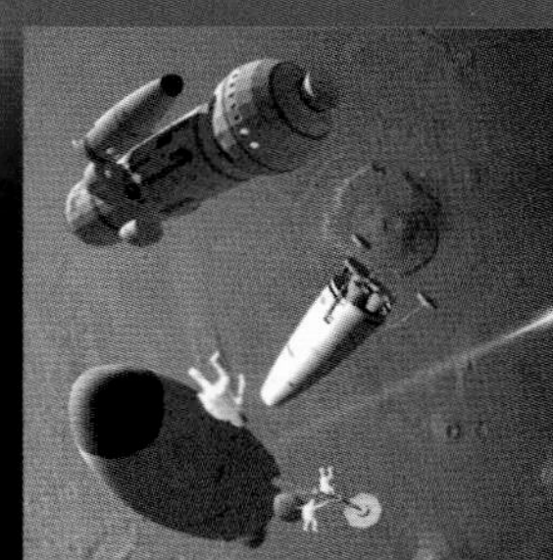

FIREFLY BOOKS

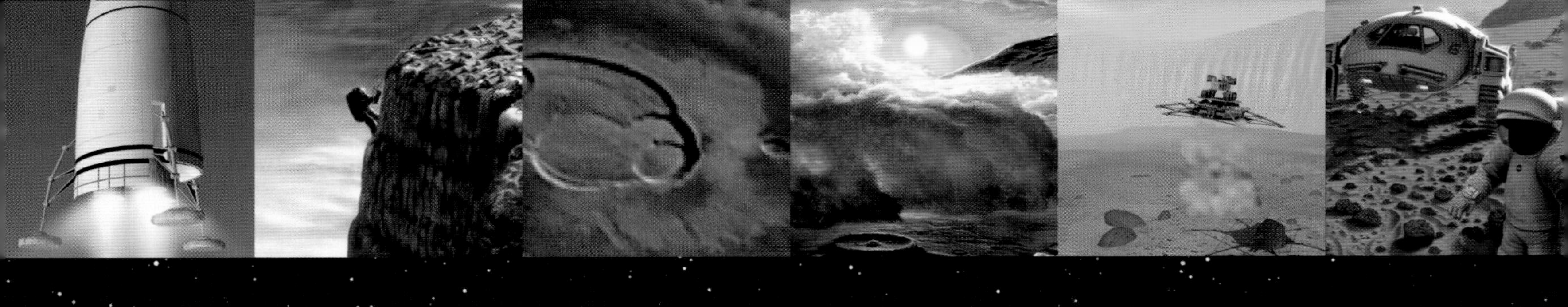

5 The Great Voyage

6 On Mars

7 Back to Earth

8 The Future

The Call of Mars

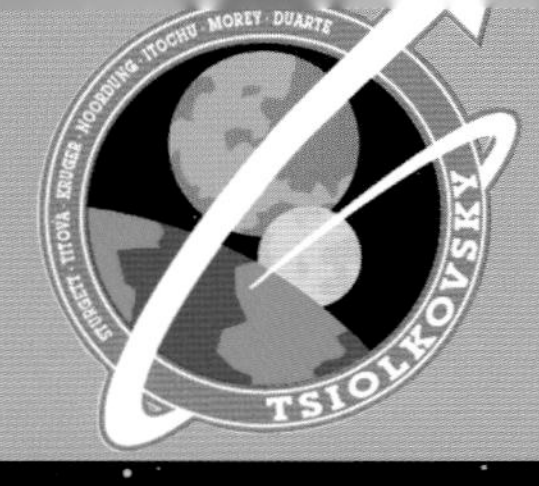

Just a small blood-red spot in the night sky, Mars is not the most spectacular object in the heavens. The Moon shines with an incomparable brightness, and two other planets, Venus and Jupiter, can be brighter. But Mars has indirectly presided over the development of humanity into a scientific and technological civilization during the last few millennia. Aside from the Sun and Moon, no other heavenly object has been as important as Mars in the history of humanity.

Mars may hold the answer to a fundamental question asked by science: Has life appeared anywhere other than on Earth? Throughout time, people have dreamed of conquest. Today, space exploration offers it to them. If the expansion of humanity is to occur beyond planet Earth, it will likely start with Mars, which could become a second Earth, another planet where men, women and children could settle, work and live.

This book is a voyage in space and time; it is fiction based on the latest scientific facts. It invites you for a voyage on the Tsiolkovski, an imaginary craft in the year 2030, with a crew of astronauts on board.

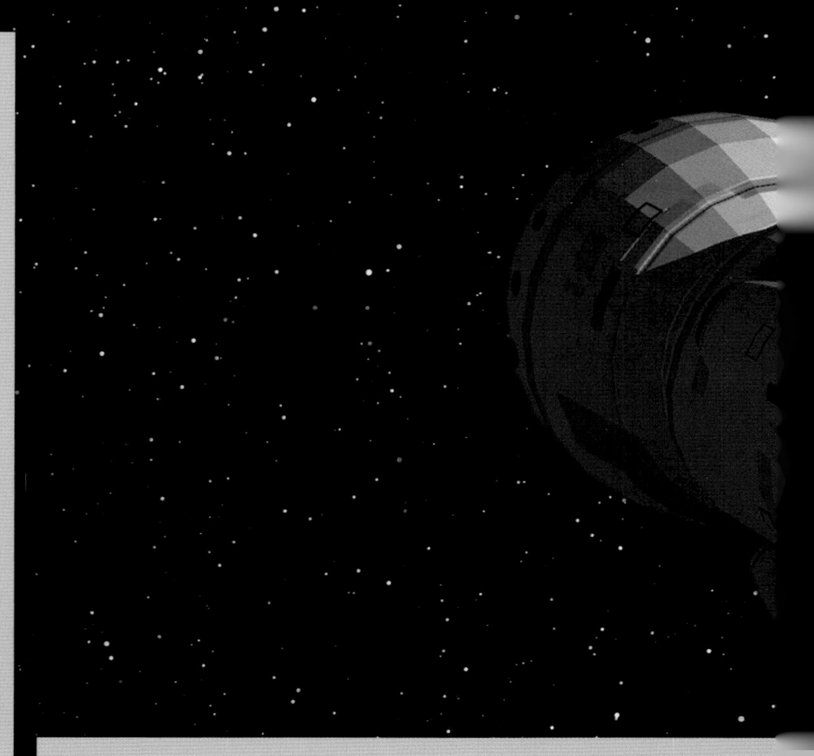

The team is international, like that of the space station occupied since the year 2000, which is the starting base for this expedition. Among the team is French astronaut Michel Morey, scientific director of the expedition. The commanding officer is an American, John Sturgett. His assistant commander, Natasha Titova, is Russian. Otto Kruger, from Germany, is the pilot of the landing craft; the leader of the Martian Camp, Jeannette Noordung, is from Canada; Nagatomo Itochu, the head of the base camp on Phobos, one of the two Martian satellites, hails from Japan; and the doctor on board is Eduardo Duarte, an American of Brazilian origin.

It has been seven decades since Neil Armstrong and Buzz Aldrin landed on the Moon. Now, several months after *Tsiolkovski's* departure, these seven people will reach Mars. Its exploration could open new perspectives for the future. The red planet shines high in the firmament of humanity's future. The dawning millennium could really be Martian!

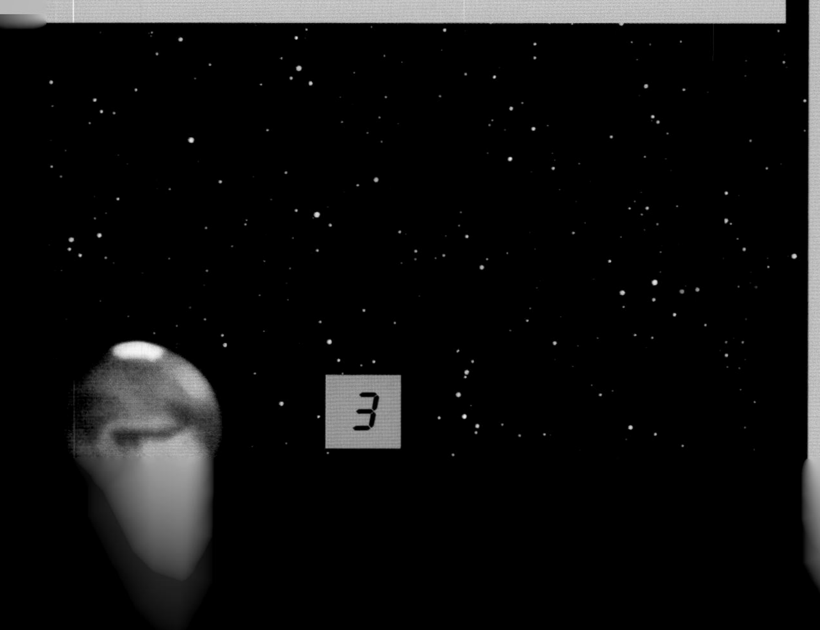

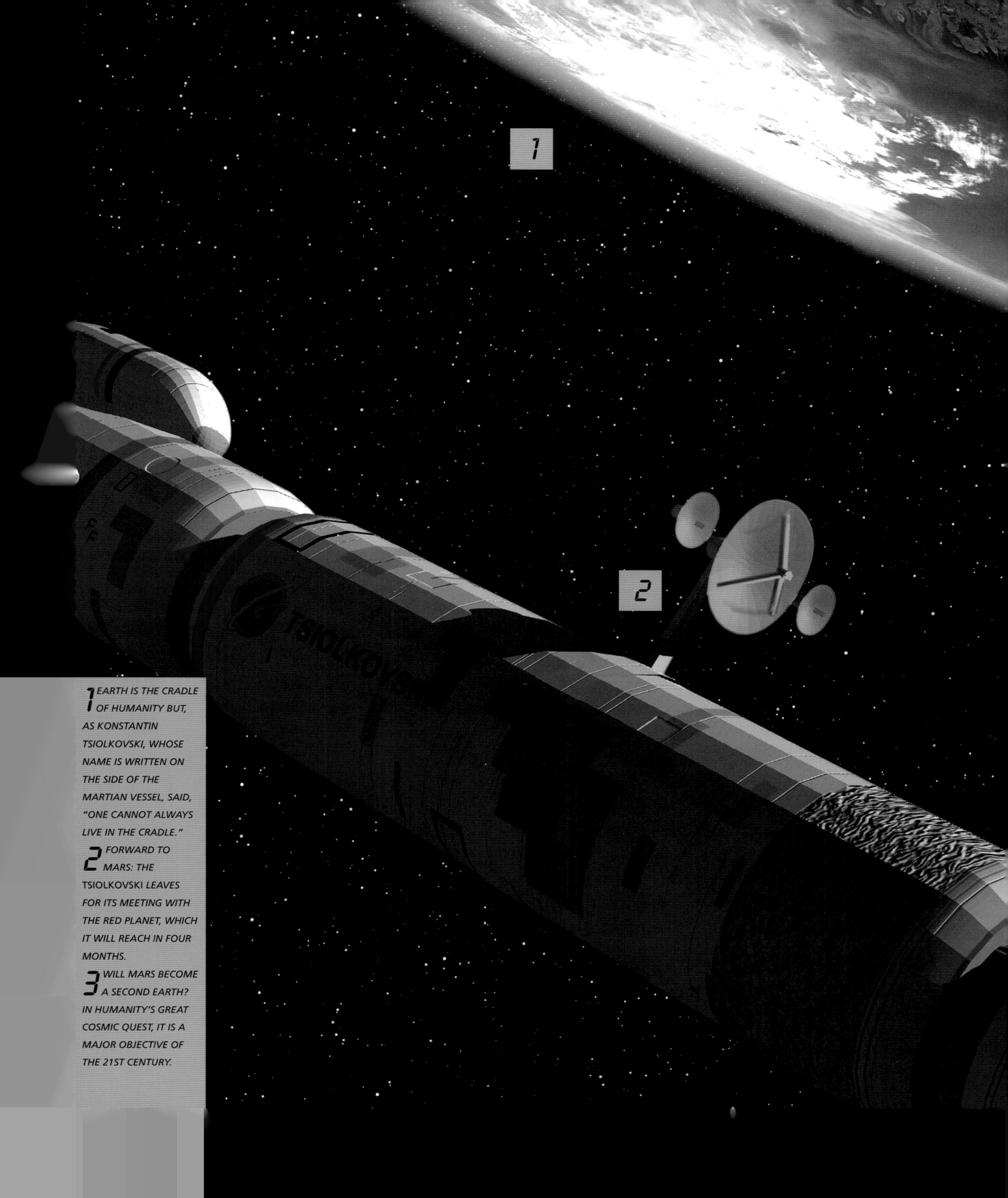

1 EARTH IS THE CRADLE OF HUMANITY BUT, AS KONSTANTIN TSIOLKOVSKI, WHOSE NAME IS WRITTEN ON THE SIDE OF THE MARTIAN VESSEL, SAID, "ONE CANNOT ALWAYS LIVE IN THE CRADLE."

2 FORWARD TO MARS: THE TSIOLKOVSKI *LEAVES FOR ITS MEETING WITH THE RED PLANET, WHICH IT WILL REACH IN FOUR MONTHS.*

3 WILL MARS BECOME A SECOND EARTH? IN HUMANITY'S GREAT COSMIC QUEST, IT IS A MAJOR OBJECTIVE OF THE 21ST CENTURY.

INTRODUCTION

A Passion that Goes Back to Antiquity

1 THE ROMANS OFTEN ASSOCIATED MARS WITH VENUS.

2 MARS, THE GOD OF WAR.

Since ancient times, Mars has drawn particular attention from those who have observed the heavens. Its behavior is stranger than that of the other planetary "wanderers" (the word "planet" comes from the Greek word for "wanderer") that move on the dome of the sky while the stars remain fixed. Its red color is unique, and its distinctiveness caused it to be associated with important gods throughout human history. For the Egyptians, Mars was Sekd-ed-ef em Khetkhet, "the star that moves backwards while traveling." And under the name of Heru-Khuti, it was one of the principal forms of the Sun god Ra, whose sanctuary was located at Heliopolis. It is even possible that the great Sphinx at Giza might have been dedicated to him.

For the Babylonians, the red planet was associated with Nergal, the star of death, which played a large role in *The Epic of Gilgamesh*. The Greeks associated it with Aries, the god of war, by whom Aphrodite bore two sons, Deimos (terror) and Phobos (fear). The planet's name as we now commonly know it is taken from the Romans; Mars was the father of Remus and Romulus, the founders of Rome.

Important for mythology, and for astrology – which started in Mesopotamia before spreading to Greece and Rome – Mars also played a major role in the development of astronomy. The philosophers of the Athenian school, including Aristotle, the principal disciple of Plato, built a model of the universe in agreement with the perfection that they thought must characterize the celestial world – and divine nature. At the center is the Earth. The Moon, Venus, Mercury, the Sun, Mars, Jupiter and Saturn, respectively, turned on crystalline spheres around the Earth.

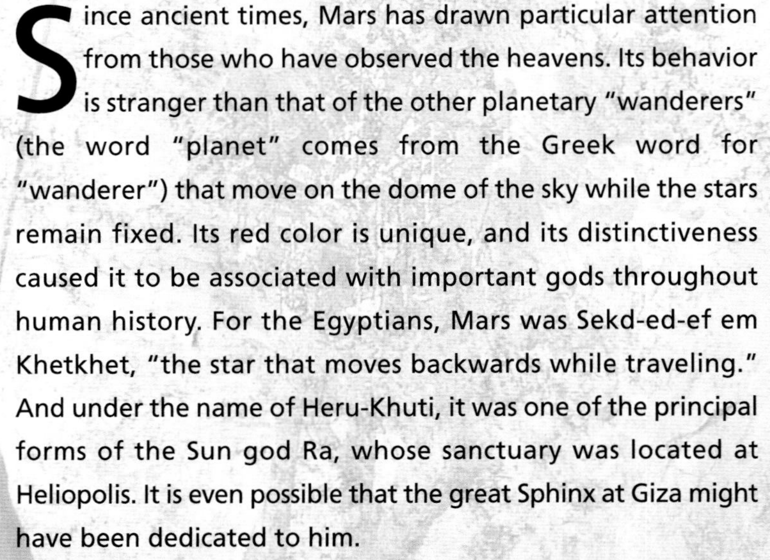

1

2

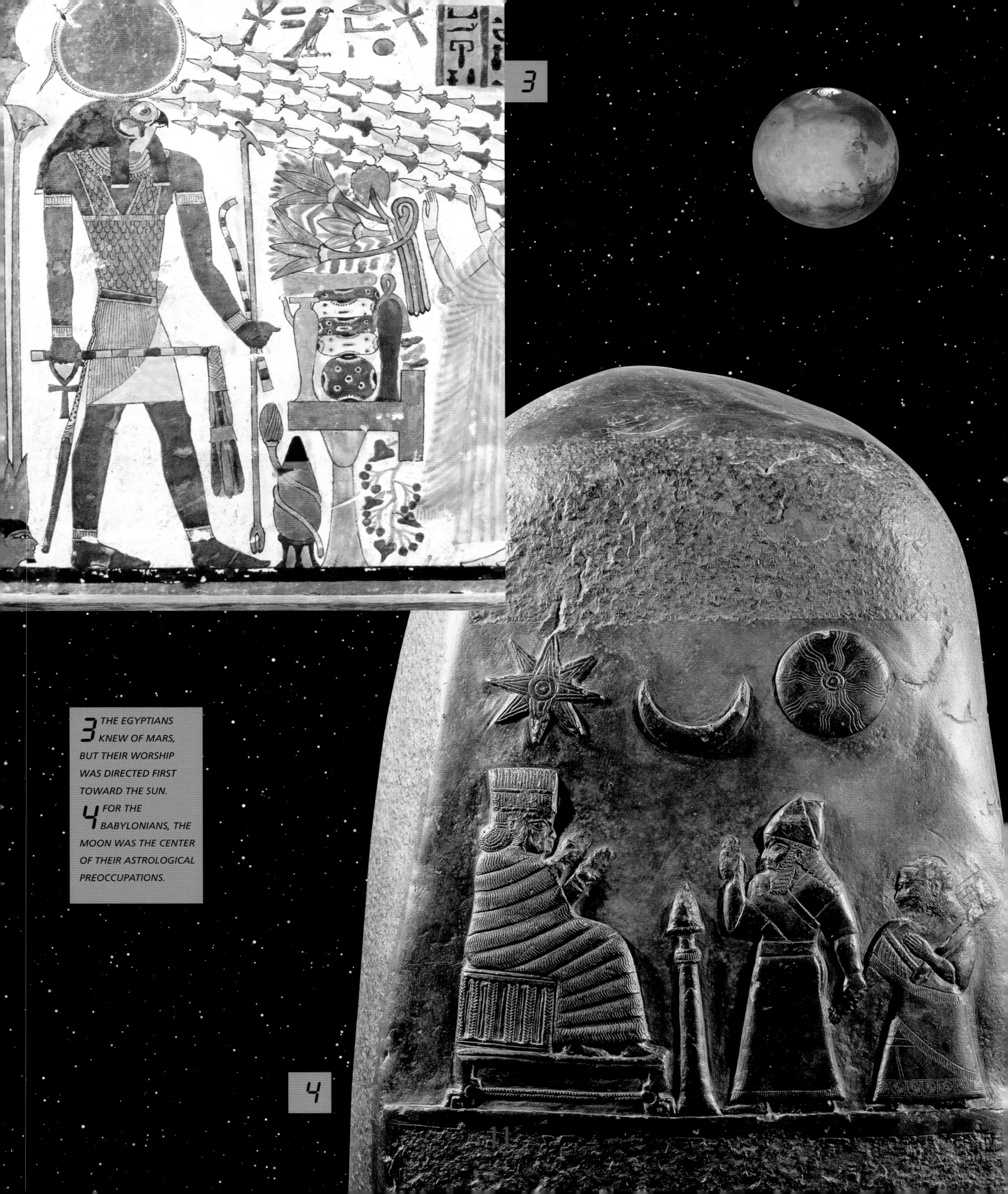

3 THE EGYPTIANS KNEW OF MARS, BUT THEIR WORSHIP WAS DIRECTED FIRST TOWARD THE SUN.
4 FOR THE BABYLONIANS, THE MOON WAS THE CENTER OF THEIR ASTROLOGICAL PREOCCUPATIONS.

1 The Mysterious Motion of Mars

The apparent motion of the five planets that were known in antiquity (Mercury, Venus, Mars, Jupiter and Saturn) was not easily reconciled with the simplicity wished for by Aristotle and his successors. To make the system work, they needed to imagine complex solutions, adding circles called epicycles on to the rotation of the crystalline spheres. Their model was brought to its highest form by Ptolemy of Alexandria at the beginning of our era and was incorporated into the Christian religion by St. Thomas Aquinas. From that point on, it became the absolute truth, the dogma that was imposed until the Renaissance.

However, despite all efforts, Mars resisted Aristotelian explanations. Its movement on the celestial dome is bizarre. Every 26 months, the red planet stops in its west-to-east movement across the sky, then turns back and draws out a loop before resuming its progress towards the east. Crystalline spheres and epicycles could not explain such a strange phenomenon. Further, if Mars turned around the Earth (as did the entire universe in this model) how could its brightness vary in such a significant way?

Jupiter and Saturn also trace out loops on the dome of the sky, but in a much less marked fashion. To explain the mysterious behavior that Mars exhibited, the leading scholars of the Renaissance needed to knock down the edifice built since Aristotle and lay

1 THE EARTH AND THE PLANETS REVOLVE AROUND THE SUN. THIS "REVOLUTIONARY" IDEA, PROPOSED IN ANTIQUITY, WAS THE KEYSTONE OF THE COPERNICAN SYSTEM.

2 HOWEVER, NICOLAUS COPERNICUS DID NOT DARE PUBLISH HIS FINDINGS UNTIL NEAR HIS DEATH.

3 PTOLEMY'S EARTH-CENTERED UNIVERSE, THE ULTIMATE DEVELOPMENT OF ARISTOTLE'S IDEAS, HAD DOMINATED PHILOSOPHY AND SCIENCE FOR 15 CENTURIES.

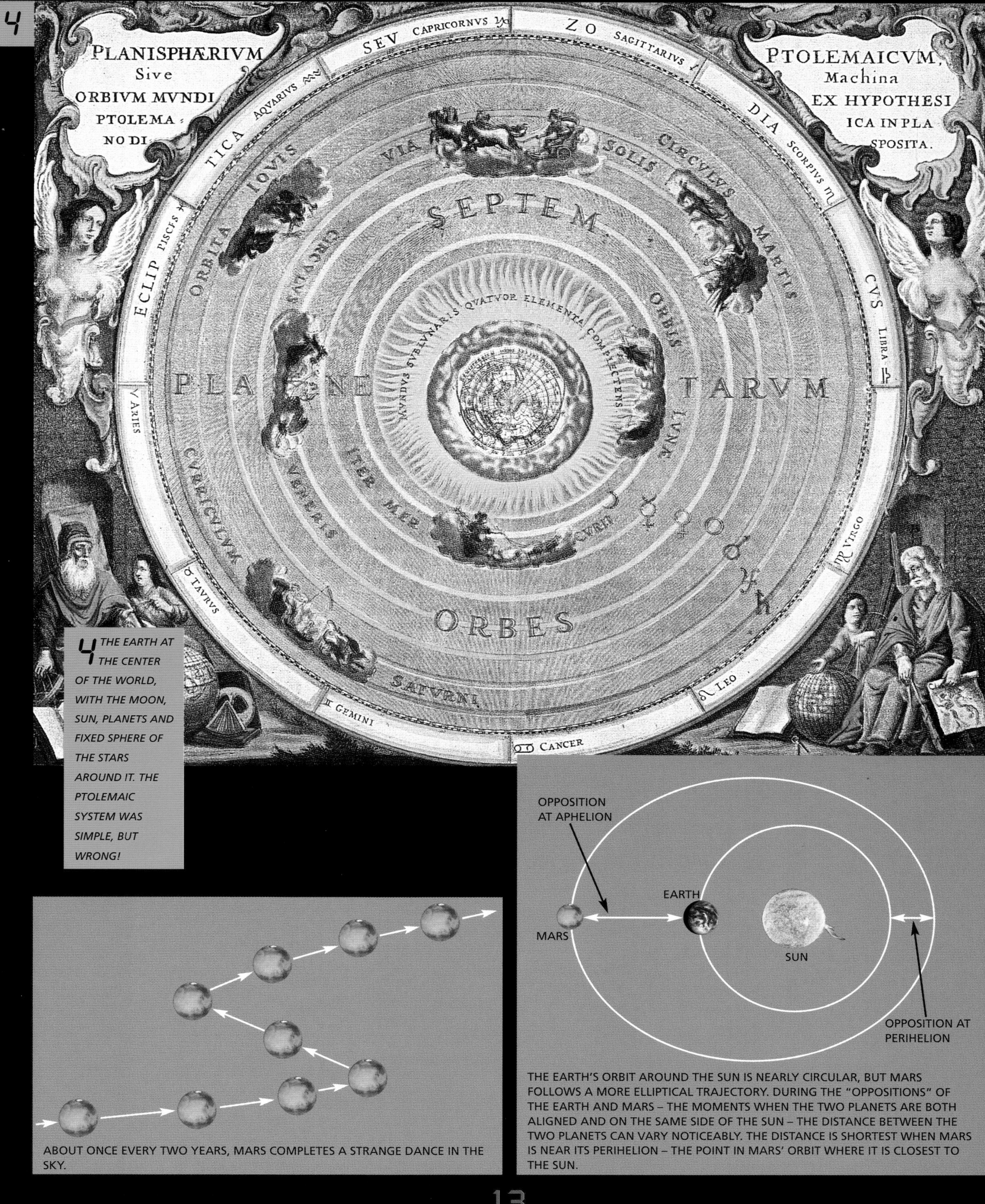

4 THE EARTH AT THE CENTER OF THE WORLD, WITH THE MOON, SUN, PLANETS AND FIXED SPHERE OF THE STARS AROUND IT. THE PTOLEMAIC SYSTEM WAS SIMPLE, BUT WRONG!

ABOUT ONCE EVERY TWO YEARS, MARS COMPLETES A STRANGE DANCE IN THE SKY.

THE EARTH'S ORBIT AROUND THE SUN IS NEARLY CIRCULAR, BUT MARS FOLLOWS A MORE ELLIPTICAL TRAJECTORY. DURING THE "OPPOSITIONS" OF THE EARTH AND MARS – THE MOMENTS WHEN THE TWO PLANETS ARE BOTH ALIGNED AND ON THE SAME SIDE OF THE SUN – THE DISTANCE BETWEEN THE TWO PLANETS CAN VARY NOTICEABLY. THE DISTANCE IS SHORTEST WHEN MARS IS NEAR ITS PERIHELION – THE POINT IN MARS' ORBIT WHERE IT IS CLOSEST TO THE SUN.

INTRODUCTION

1 The Origins of the Scientific Revolution

1 TYCHO BRAHE CONSTRUCTED THE LARGEST NAKED-EYE OBSERVATORY, AND JOHANNES KEPLER DISCOVERED THAT THE PTOLEMAIC SYSTEM DID NOT EXPLAIN THE UNUSUAL PATH OF MARS.

1

In the liturgy of the Middle Ages, inherited from Aristotle and Ptolemy, Mars, like the Moon and the other planets, was a perfect body formed of "quintessence," or the "fifth element," which belonged to the divine world. The Earth, on the other hand, was a "corrupted" world, made up of four elements: air, fire, earth and water. It took until the middle of the 16th century and Nicolaus Copernicus for this model to be challenged. Copernicus returned to an idea proposed in antiquity but rejected as contrary to common sense: It was the Sun that was at the center of the universe, and the five planets turned around it (only the Moon turned around the Earth, which was a planet like the others). Copernicus only dared to publish these revolutionary ideas at the end of his life in 1543, in a work entitled *De Revolutionibus Orbium Coelestium*.

However, it was Johannes Kepler who received the honor for demonstrating that Copernicus was right ... thanks to Mars! In 1609, Kepler came to a surprising conclusion. Based on his calculations intended to explain the strange trajectory of the red planet, the only means of explaining the very precise observations obtained by Tycho Brahe was to suppose that Mars did not trace out a circle around Earth, but an ellipse around the Sun. The Earth was not, therefore, at the center the universe. This fact was supported the following year by Galileo Galilei, who pointed the first astronomical telescope at the sky and discovered four satellites around Jupiter. It looked like the center of a small solar system. Galileo also observed mountains and craters on the Moon, showing it was not a perfect body.

The fundamental discovery of Kepler was the origin of Isaac Newton's calculations leading to the universal law of gravitation and the fundamental law of dynamics – meaning it is at the foundation of modern science and laws of the universe. Mars is therefore the indirect origin of the scientific and technological developments that have marked the last three centuries and that have changed our conception of nature, the universe and our daily life. Without Mars and its strange celestial movements, contemporary civilization might be very different.

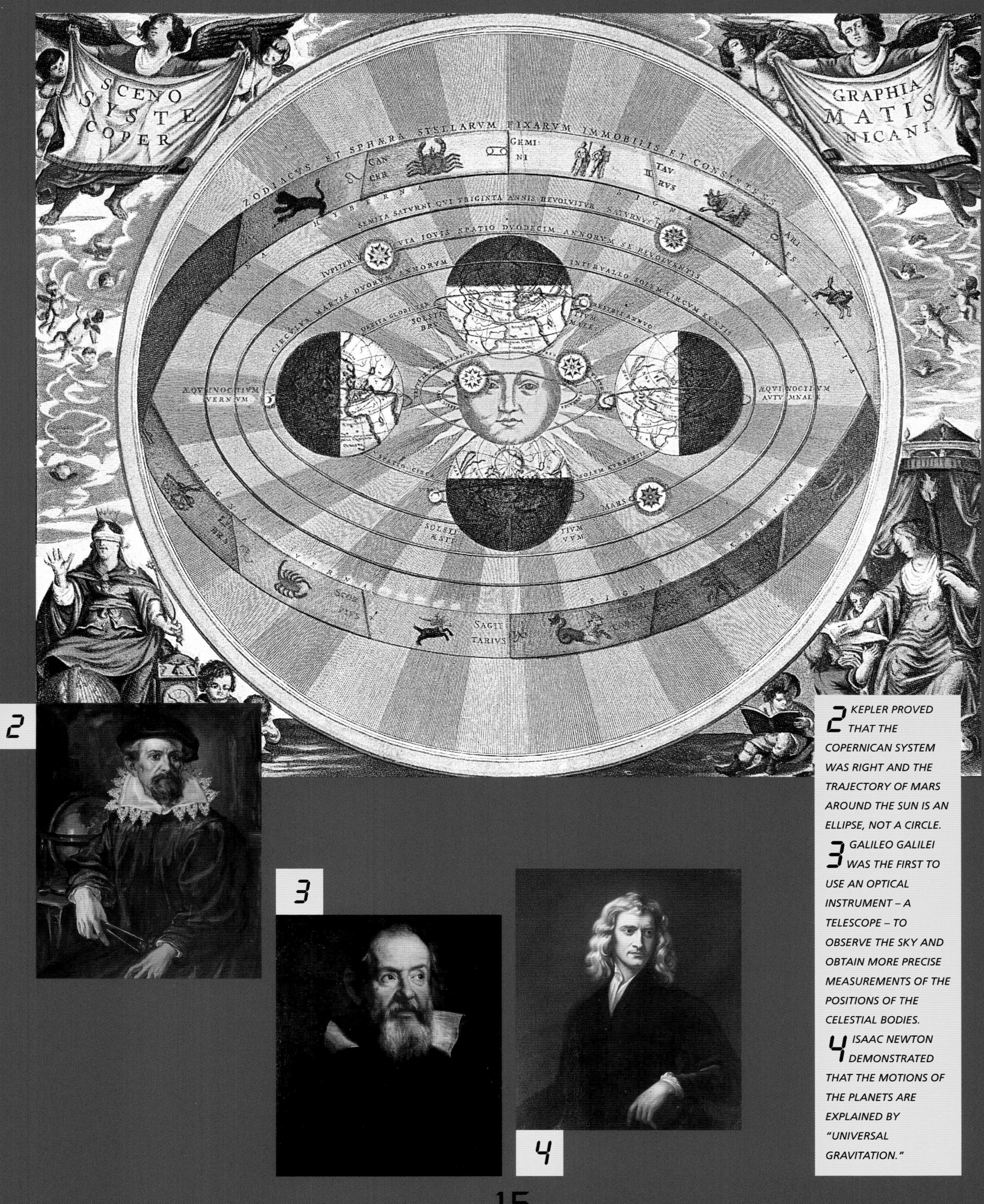

2 KEPLER PROVED THAT THE COPERNICAN SYSTEM WAS RIGHT AND THE TRAJECTORY OF MARS AROUND THE SUN IS AN ELLIPSE, NOT A CIRCLE.

3 GALILEO GALILEI WAS THE FIRST TO USE AN OPTICAL INSTRUMENT – A TELESCOPE – TO OBSERVE THE SKY AND OBTAIN MORE PRECISE MEASUREMENTS OF THE POSITIONS OF THE CELESTIAL BODIES.

4 ISAAC NEWTON DEMONSTRATED THAT THE MOTIONS OF THE PLANETS ARE EXPLAINED BY "UNIVERSAL GRAVITATION."

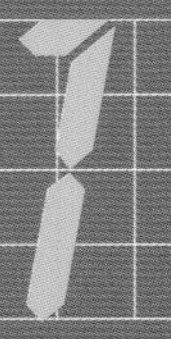

The Rhythm of Oppositions

In 1659, Christian Huygens took advantage of an "opposition" of the Earth and Mars to observe the red planet. An opposition occurs when two planets, following their paths, are aligned with the Sun and are on the same side of the Sun at the same time. At opposition, the planets are closest together. That's when the small, reddish Martian globe, fuzzy and changing, reveals a few more details in binoculars and telescopes. Huygens observed formations on the surface of the red planet for the first time (notably, he sketched the formation that was to be named Syrtis Major), and measured its rotational period, which is a little more than 24 hours. Mars truly seemed like another Earth.

The oppositions of Mars repeat every 26 months and that's the rhythm that observations of the red planet follow. Further, some oppositions are more favorable than others. Mars comes particularly close to the Earth when opposition occurs while Mars is at "perihelion" – the point in Mars' orbit closest to the Sun. Opposition occurs at perihelion every 13 or 15 years; at this time the red planet is only 35 million miles (56 million km) from the Earth. Astronomers take considerable effort to get the maximum advantage from these exceptional events.

That was the case in 1672, when astronomer Jean Richer went on an expedition to Cayenne, French Guiana, to observe the planet from a site far removed from Paris. He used a method known as "triangulation" – an object (in this case, Mars) is sighted at the same moment from two different points on Earth with a known distance between them, and a "triangle" is created by, in effect, "connecting the dots." Taking the angles and direct measurements into account, he was able to approximate the distance to Mars at the moment of opposition. Based on obtaining this measurement, it was then possible to determine all distances in the solar system, and, in particular, the average distance from the Earth to the Sun, known as the astronomical unit, or AU. The calculations were performed at the Paris Observatory by Giovanni Domenico Cassini. He came to a fairly accurate value for the astronomical unit – about 80 million miles (130 million km). (Today it is known that this unit is actually somewhat larger: 92.9 million miles/ 149.5 million km.)

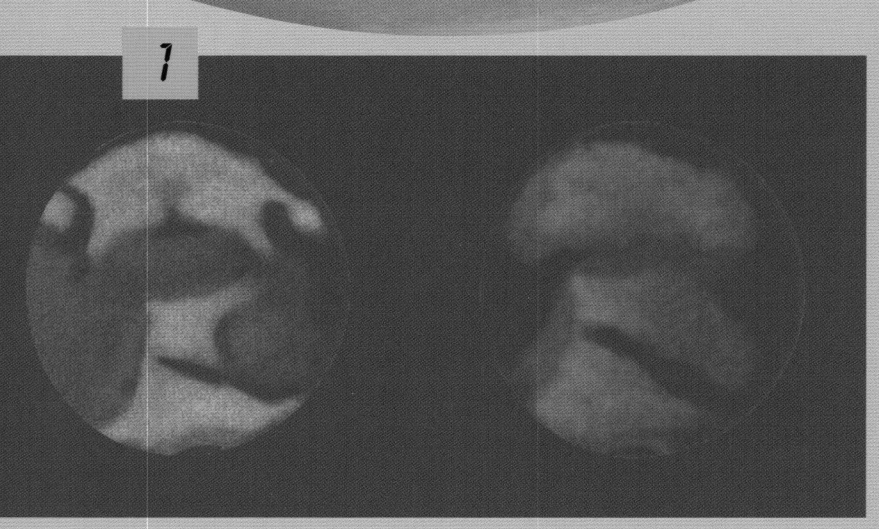

Date	Apparent Dimension of Martian Disk	Earth-Mars Distance in millions of miles (millions of km)
28/08/2003	25.1 arc sec.	34.7 (55.8) *opposition at perihelion
7/11/2005	19.8 arc sec.	43.7 (70.3)
28/12/2007	15.5 arc sec.	55.7 (89.7)
29/01/2010	14 arc sec.	61.7 (99.3)
3/03/2012	14 arc sec.	62.6 (100.8)
8/04/2014	15.1 arc sec.	57.7 (92.9)
22/05/2016	18.4 arc sec.	47.3 (76.1)
27/07/2018	24.1 arc sec.	35.9 (57.7) *opposition at perihelion
13/10/2020	22.3 arc sec.	39.0 (62.7)
8/12/2022	16.9 arc sec.	51.1 (82.3)
16/01/2025	14.4 arc sec.	59.8 (96.2)
19/02/2027	13.8 arc sec.	63.0 (101.4)
25/03/2029	14.4 arc sec.	60.3 (97.1)
4/05/2031	16.9 arc sec.	51.9 (83.6)
27/06/2033	22 arc sec.	39.7 (63.9)
15/09/2035	24.5 arc sec.	35.5 (57.1) *opposition at perihelion

DATES AND PROPERTIES OF UPCOMING MARS OPPOSITIONS.

1 FUZZY FORMATIONS ON THE SMALL MARTIAN DISK WERE DISTINGUISHED DURING OPPOSITIONS OF EARTH AND MARS.

2 THE MAJOR ASTRONOMERS OF THE 17TH CENTURY TOOK ADVANTAGE OF NEW MATHEMATICAL CALCULATIONS BASED ON THE KNOWLEDGE OF OPPOSITIONS, INCLUDING CASSINI, THE FOUNDER OF THE PARIS OBSERVATORY ...

3 ... AND CHRISTIAN HUYGENS, WHO DISCOVERED THAT A DAY ON MARS LASTS A LITTLE LONGER THAN A DAY ON EARTH ...

4 ... AND WILLIAM HERSCHEL, WHO DISCOVERED THE PLANET URANUS.

The Canals of Mars

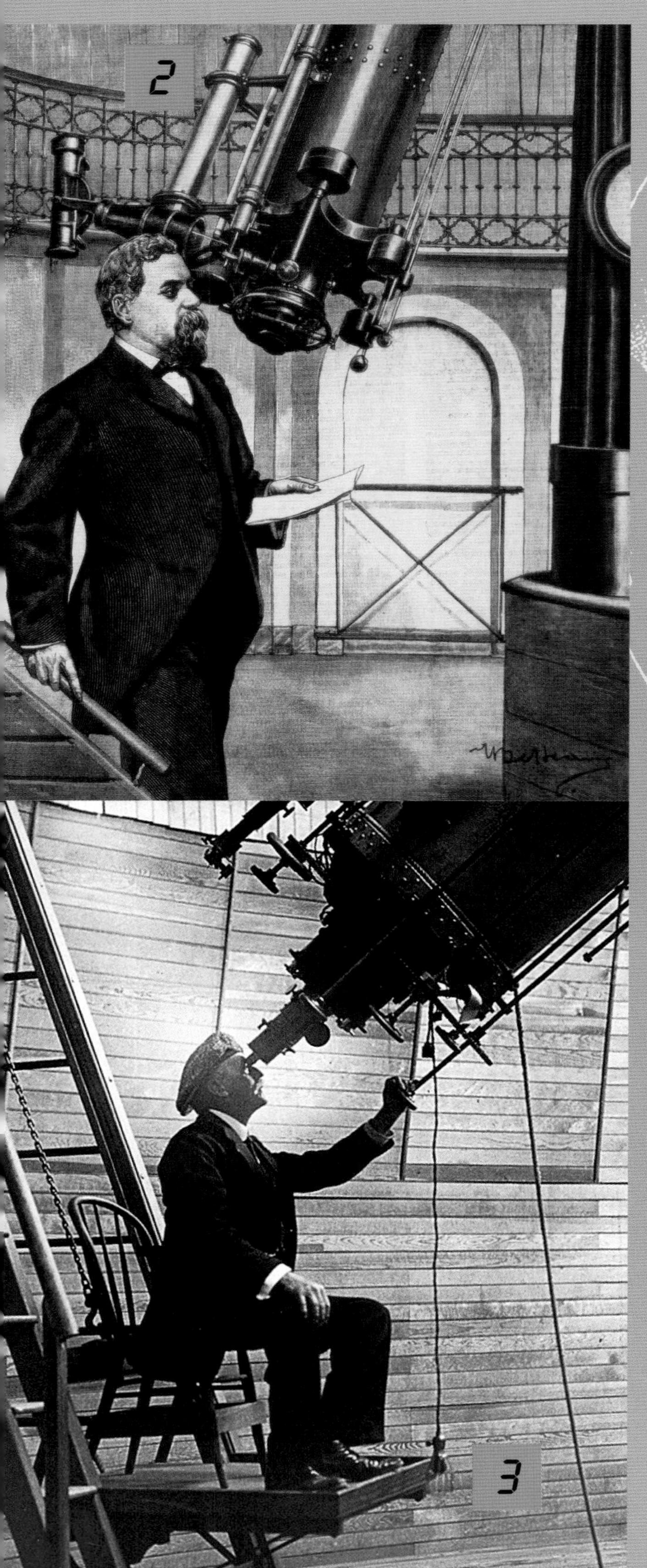

The opposition of Mars that took place in 1877 was exceptional. Many observatories aimed their telescopes at the red planet, and the results were spectacular. Asaph Hall discovered the two small satellites of Mars – Deimos and Phobos – whose existence had been postulated by Johannes Kepler (and also by Jonathan Swift in *Gulliver's Travels*). But it was Giovanni Virginio Schiaparelli who observed what he called *canali* ("rivers" in Italian) on Mars; however, *canali* was translated as "canals," implying an artificial construction. Well-known astronomers, including Stanley Williams in England, H.C. Wilson in the United States, and Henri Perrotin and Louis Thollon at the Nice Observatory in France, confirmed the views of Schiaparelli.

A wealthy amateur astronomer, Percival Lowell (1855–1916), became passionate about this idea and committed his money and time to the "canals of Mars." He had a large telescope built in Flagstaff, Arizona, dedicated to the study of the red planet. With each succeeding observation, he improved his map of the canals and developed a theory that wasn't particularly scientific: The "canals" were used to distribute water that had become rare on a planet that was drying out. Of course, these were not real canals, as such. What could be seen from Earth were bands of vegetation, just like cultivated zones bordering the Nile in Egypt. That meant that Mars had an ancient civilization whose inhabitants were fighting for survival – a great story! And it was a story that, at the time, was taken on the whole to be real. It wasn't until 1909 that the idea of Martian canals was completely abandoned by astronomers.

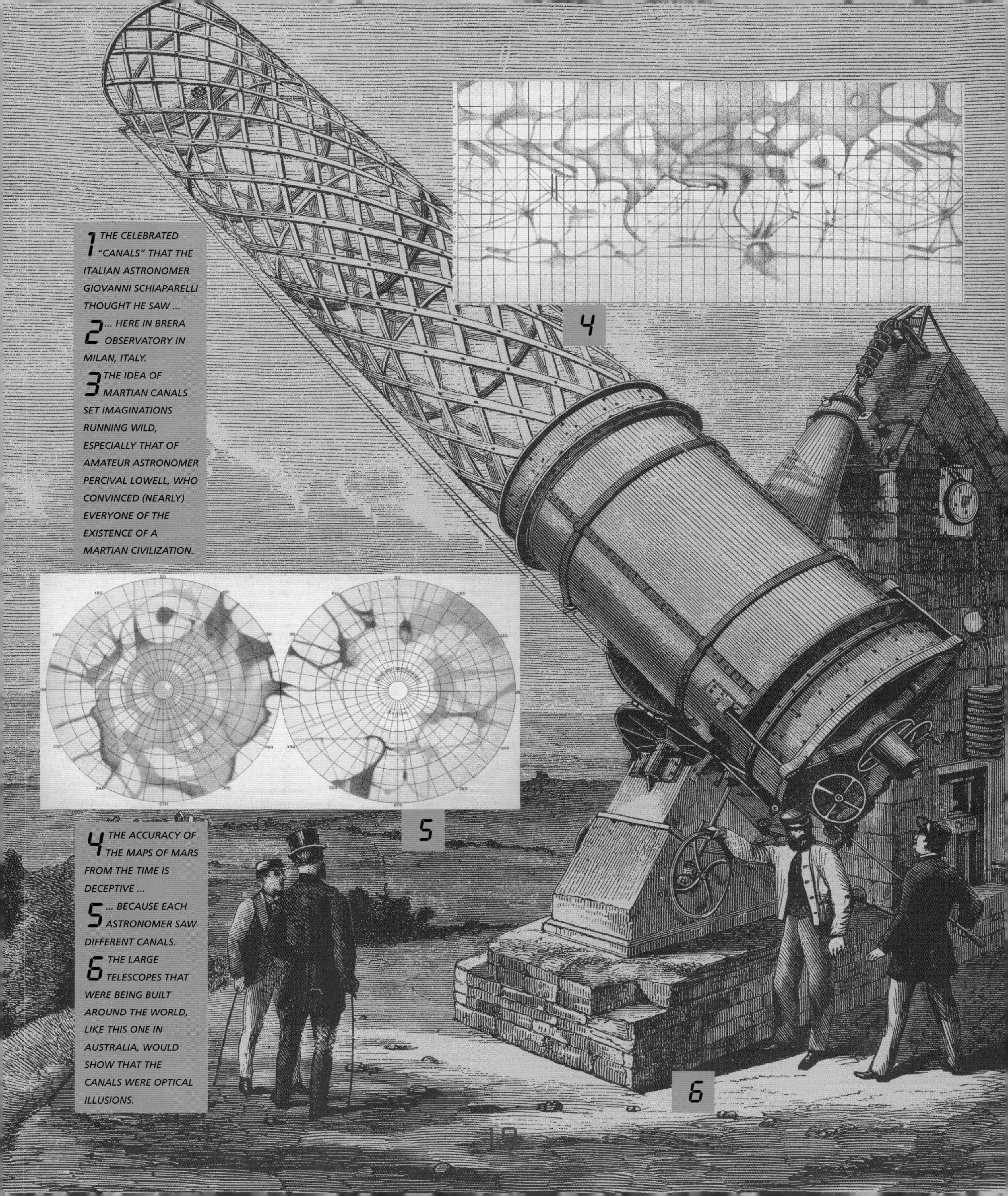

1 THE CELEBRATED "CANALS" THAT THE ITALIAN ASTRONOMER GIOVANNI SCHIAPARELLI THOUGHT HE SAW ...

2 ... HERE IN BRERA OBSERVATORY IN MILAN, ITALY.

3 THE IDEA OF MARTIAN CANALS SET IMAGINATIONS RUNNING WILD, ESPECIALLY THAT OF AMATEUR ASTRONOMER PERCIVAL LOWELL, WHO CONVINCED (NEARLY) EVERYONE OF THE EXISTENCE OF A MARTIAN CIVILIZATION.

4 THE ACCURACY OF THE MAPS OF MARS FROM THE TIME IS DECEPTIVE ...

5 ... BECAUSE EACH ASTRONOMER SAW DIFFERENT CANALS.

6 THE LARGE TELESCOPES THAT WERE BEING BUILT AROUND THE WORLD, LIKE THIS ONE IN AUSTRALIA, WOULD SHOW THAT THE CANALS WERE OPTICAL ILLUSIONS.

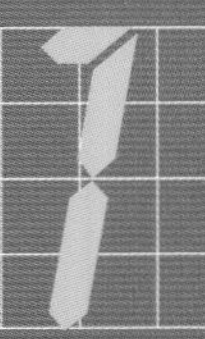

Mars in Literature

Jules Verne (1828–1905), author of *From the Earth to the Moon and Round the Moon*, is one of the greatest masters of science fiction writing. The writer was, of course, mistaken about many points. A cannon is not a good way to send humans into space (the poor astronauts would be crushed upon launching), and weight does not disappear only at the point where the attractions of the Earth and the Moon balance out (weightlessness is the rule aboard a spacecraft only once the rockets stop). But the enthusiasm in all of Verne's books is compelling, and it inspired some of the great pioneers of space adventure, like Konstantin Tsiolkovski and Wernher von Braun, the father of modern rocketry.

Jules Verne was not interested in Mars, but the red planet became an important source of inspiration for 20th-century science fiction. *The War of the Worlds*, by the great English writer H.G. Wells, was published in 1898. It featured pitiless Martians who wanted to destroy humanity.

While we know today that this image of brutal Martian invaders is false, it did a great deal to popularize the idea that Martians could exist.

The American author Edgar Rice Burroughs also has an important place in spreading that idea. Inspired by Giovanni Schiaparelli's and Percival Lowell's visions of Martian canals, he wrote a series of very popular science fiction novels. The hero in these books, John Carter, was miraculously transported to Mars and aided the inhabitants in surviving on an inexorably drying planet. *A Princess of Mars* and its

1 *H.G. WELLS SCARED THE WORLD WITH THIS MARTIAN FROM THE* WAR OF THE WORLDS ...

sequels were an immense success in the 1920s and were the inspiration of many illustrated pulp fiction tales.

Mars also inspired other works of science fiction, notably *The Martian Chronicles* by Ray Bradbury, published just after World War II. More recently, novels have sought to give life to the idea of the colonization of Mars put forth by trailblazers such as Robert Zubrin. The trilogy by Kim Stanley Robinson, *Red Mars, Blue Mars* and *Green Mars*, falls in that category.

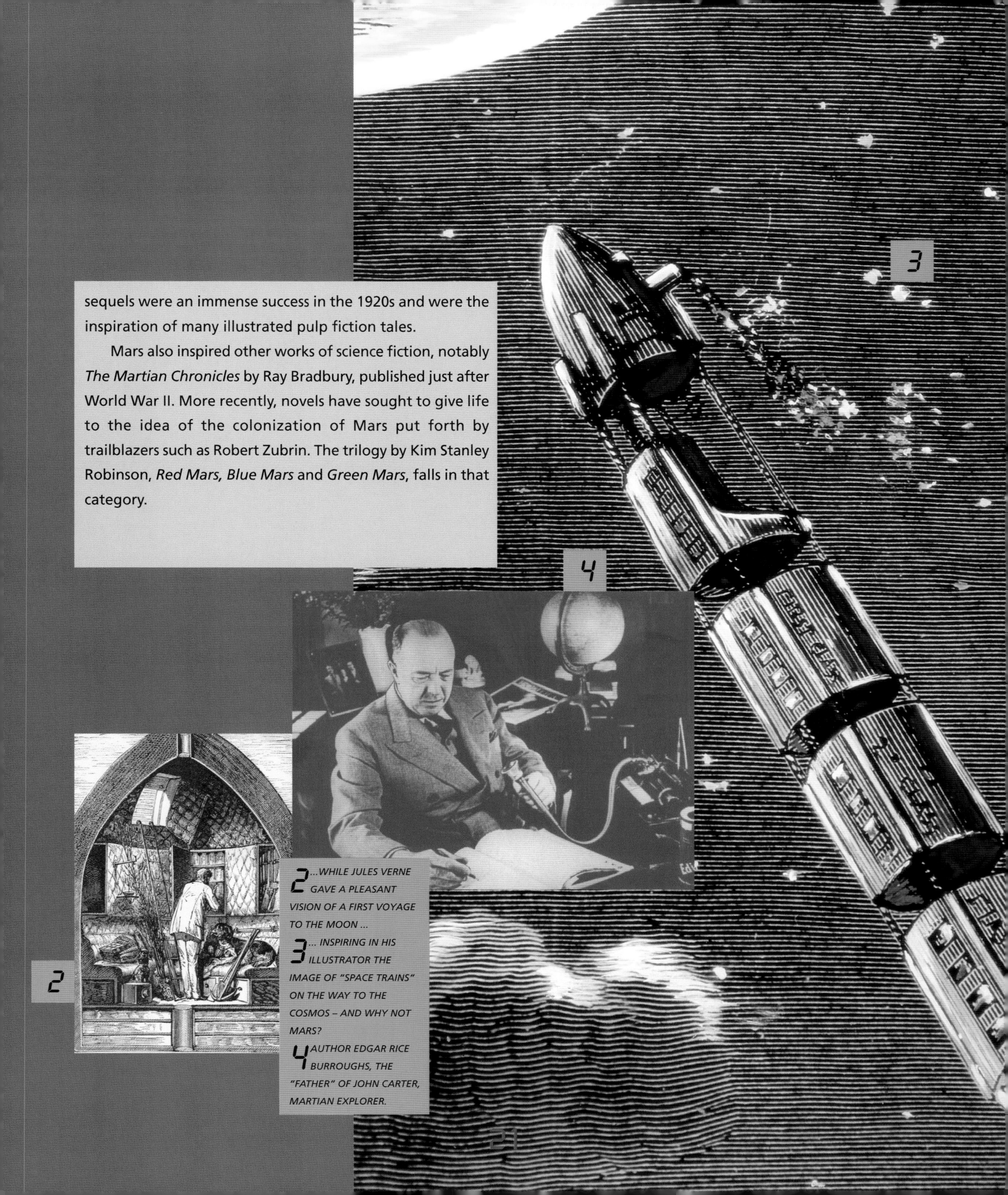

2 ...WHILE JULES VERNE GAVE A PLEASANT VISION OF A FIRST VOYAGE TO THE MOON ...

3 ... INSPIRING IN HIS ILLUSTRATOR THE IMAGE OF "SPACE TRAINS" ON THE WAY TO THE COSMOS – AND WHY NOT MARS?

4 AUTHOR EDGAR RICE BURROUGHS, THE "FATHER" OF JOHN CARTER, MARTIAN EXPLORER.

INTRODUCTION

1 Mars in the Movies

If science fiction novels have played a large role in the preparation of the public consciousness for the coming of the Space Age and the conquest of Mars, movies and television have also made big contributions.

Beginning in 1902, Georges Méliès transformed the emerging art of cinema into a fantastic medium of expression for science fiction, with his masterpiece *A Trip to the Moon (Le Voyage dans la Lune)*. In 1929 the great German director Fritz Lang filmed *The Woman in the Moon*; his technical advisor, Herman Oberth, was one of the great trailblazers of space exploration.

Mars became the subject of many films, including director Paul Verhoeven's *Total Recall*, which depicted a Martian colony, and Tim Burton's *Mars Attacks*, a hilarious portrayal of fiendish Martians.

1

2

1 A "REALISTIC" IMAGE OF SPACE *FLIGHT FROM* 1929: THE WOMAN IN THE MOON *BY FRITZ LANG.*

2 WERNHER VON *BRAUN INSPIRED* CONQUEST OF SPACE, *TWO YEARS BEFORE SPUTNIK.*

3

3 *THE SPIRITUAL DIMENSION OF THE COSMOS* – 2001: A SPACE ODYSSEY *BY STANLEY KUBRICK IN 1969.*

4

4 THANKS TO TELEVISION AND *THEN MOVIES, THE "STAR TREK GENERATION" WANTS TO GO TO SPACE FASTER THAN EVER BEFORE.*

5 *WHAT IF MARTIANS WERE VERY MEAN?* MARS ATTACKS.

Cinema is not limited by technology; anything can be imagined. Screenwriters and directors take their heroes beyond the solar system, and around other stars – even to other galaxies. If the required technology to make such travel possible doesn't exist, that's no obstacle. Authors like Arthur C. Clarke and Isaac Asimov solved the problem by imagining "jumps into hyperspace."

Interstellar adventure tales and "space operas" have met with worldwide success, including *Forbidden Planet*, the *Star Wars* movies by George Lucas, the Alien series started by Ridley Scott, *Star Trek* (adapted from a cult TV series), *Independence Day* and so on. Some films pose serious and interesting questions. That is certainly true of *2001: A Space Odyssey* by Stanley Kubrick;, based on the novel by Arthur C. Clarke. It asks profound questions about humanity's place in the universe. And humanity's place in the universe is a question undeniably at the heart of Martian exploration.

5

6 MISSION TO MARS: *MOVIE INTEREST IN MARS CONTINUE INTO THE 21ST CENTURY.*

MISSION TO MARS

6

BRIAN DE PALMA

1 Mars: A Continuing Mystery

The controversy that surrounded the Martian canals showed that despite technical progress achieved in the design of instruments, until the beginning of the 20th century, observation of the red planet amounted as much to art as to science. Even now, binoculars and telescopes have limited magnification and are often located at sites where our humid and turbulent atmosphere disrupts the work.

Before the 20th century, the eye that watched and the hand that drew remained the principal means of describing celestial objects, and that left an available void to be filled by the imagination of astronomers.

The great distance between Earth and Mars added to the lack of precision in observations. In the best cases (during "perihelion" oppositions) it is still 35 million miles (56 million km) from the Earth – 140 times farther than the Moon. It appears no larger than a plate observed from 8,200 feet (2,500 m) away!

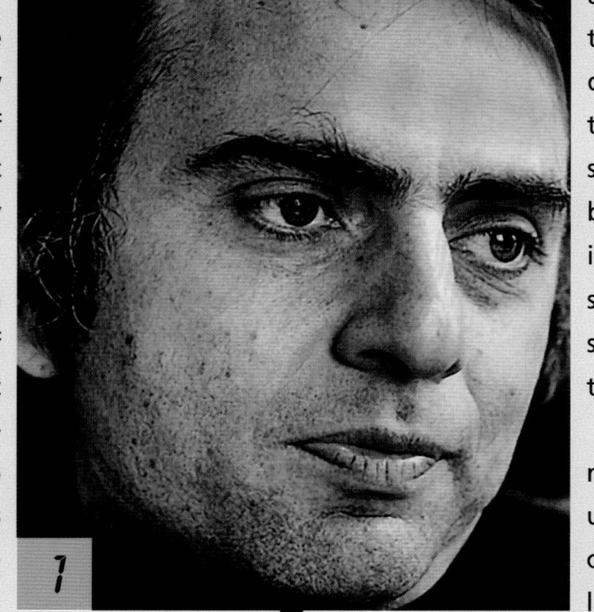

The observations are even more difficult because, during oppositions, the Sun is at the zenith above the side of Mars visible from the Earth, and the contrasts are therefore reduced. Because of this, all the maps made of Mars were different. Only the polar ice caps – seen by Giovanni Domenico Cassini in 1666 – and a few of the larger features (like Syrtis Major) seemed to be permanent fixtures on the Martian surface. Most of the other features seemed to change shape or color from one opposition to the next, when they weren't simply overshadowed – like the Martian canals that practically disappeared from the astronomical landscape after the death of Percival Lowell in 1916.

Astronomical study of Mars in the first half of the 20th century raised more questions than it answered. Were the apparent dark areas oceans, and were the light areas continents? Was the darkening of the planet that seemed to spread from the poles at certain seasons caused by vegetation benefiting from the melting of polar ice? What were the clouds that sometimes seemed to mask its surface? And what was the bluish veil that surrounded the planet?

Despite advances in astronomy, most of these questions remained unanswered until 1957, the beginning of the Space Age. Until then, the idea, largely inspired by Lowell, that Mars might be covered with vegetation remained popular among astronomers. Of course, there is no longer any question of dying civilizations and artificial canals. We now know that Mars is colder than we had thought, and its atmosphere is poor in oxygen and water vapor. But the idea of discovering some form of life there is tenacious. And only very close observations, with spacecraft, can provide definite knowledge.

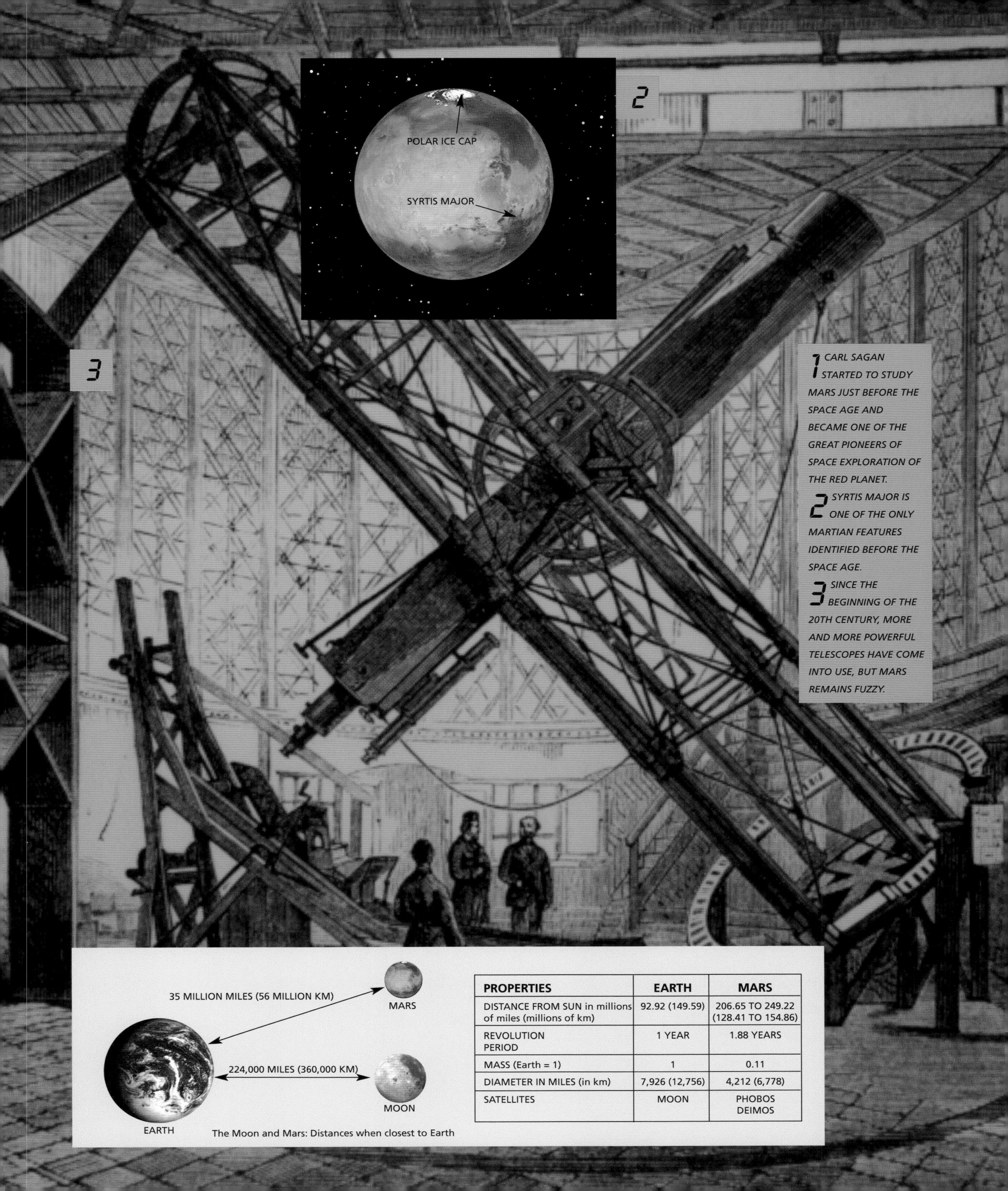

The Moon and Mars: Distances when closest to Earth

1 CARL SAGAN STARTED TO STUDY MARS JUST BEFORE THE SPACE AGE AND BECAME ONE OF THE GREAT PIONEERS OF SPACE EXPLORATION OF THE RED PLANET.

2 SYRTIS MAJOR IS ONE OF THE ONLY MARTIAN FEATURES IDENTIFIED BEFORE THE SPACE AGE.

3 SINCE THE BEGINNING OF THE 20TH CENTURY, MORE AND MORE POWERFUL TELESCOPES HAVE COME INTO USE, BUT MARS REMAINS FUZZY.

PROPERTIES	EARTH	MARS
DISTANCE FROM SUN in millions of miles (millions of km)	92.92 (149.59)	206.65 TO 249.22 (128.41 TO 154.86)
REVOLUTION PERIOD	1 YEAR	1.88 YEARS
MASS (Earth = 1)	1	0.11
DIAMETER IN MILES (in km)	7,926 (12,756)	4,212 (6,778)
SATELLITES	MOON	PHOBOS DEIMOS

2 The Long Road to Mars: The First *Mariner* Missions

In October 1957, the first artificial satellite, *Sputnik*, circled the Earth. From that point on, scientists dreamed of using spacecraft to get a closer look at the planets, which remained shrouded with an air of mystery that could not be pierced from a distance of hundreds of millions of miles.

Venus and Mars are our closest and most accessible planets, making them of special interest to astronomers. Despite the relative proximity, however, organizing an expedition to Venus or Mars is not easy.

First, the spacecraft must reach a speed much greater than that of *Sputnik* – about 26,000 mph (42,000 km/h), compared to 17,500 mph (28,000 km/h) for an artificial satellite near the Earth. Second, the spacecraft must be guided with sufficient precision for it to pass near Venus or Mars after a long journey – about 190 million miles (300 million km) for the evening star, covered in about three months, and twice that for the red planet, which is only reached after six to nine months. Finally, radio communication must also cross these phenomenal distances.

As well, the optimum opportunities to send a spacecraft toward Venus or Mars are rare. To be economical, the launches must take place a little before the planet's opposition with the Earth, which occur every 16 months for Venus and every 26 months for Mars. In other words, if the "window of opportunity" for Mars is missed, then it's necessary to wait two years to try again.

1

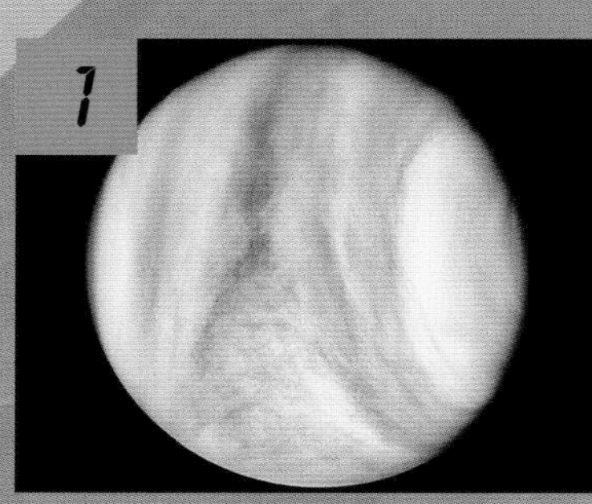

2

3

4

But the hunger for discovery aroused by the planets, and in particular the mysterious Mars, was stronger than the obstacles. Only a few years after *Sputnik*, the United States and USSR began launching interplanetary probes.

In 1961, Venus was the Soviets' first planetary target; but they did not meet with success. Mars' turn came in 1962. The Soviets made three attempts, but only one succeeded: *Mars 1* left the Earth on November 1, but broke down after a few months and was inactive when it passed near Mars on June 19, 1963.

In the same year, the U.S. had better luck with Venus. *Mariner 2* left Cape Canaveral, Florida, on August 27, 1962, and passed about 17,000 miles (27,000 km) from Venus in December. The simple flyby was very short: *Mariner 2* could only study the evening star for 42 minutes before continuing its journey across the solar system. But even such a brief observation was sufficient to overturn established knowledge. *Mariner 2* revealed that under its impenetrable clouds, Venus is a burning inferno, with an atmosphere a hundred times more dense than the Earth's and a temperature near 930°F (500°C).

The exploration of Mars by interplanetary probes, which resumed two years after *Mars 1*, would be the source of even more surprising discoveries. *Mariner 4* passed 6,120 miles (9,850 km) from the red planet on July 14, 1965, and took 22 pictures.

Their retransmission toward the distant Earth took about 10 days. The technical achievement was remarkable, but for anyone who might have been influenced by Percival Lowell, the disappointment would have been great. The photographs showed a surface covered with craters, drawing comparisons not to a new Earth but to another Moon. The images from *Mariner 4* seemed to establish that the red planet, unlike the Earth, had been inactive since the end of the great asteroid bombardment that similarly shaped the lunar surface during the early days of the solar system, 3.8 billion years ago.

One successful experiment involved the analysis of the manner in which the signals from *Mariner 4* weakened before disappearing when the spacecraft passed behind Mars. During this phenomenon, called an "occultation," the final signals pass through the planet's atmosphere, providing precious information.

The air on Mars is formed predominantly of carbon dioxide. The pressure at the surface of the red planet was found to be very low, on average, about 6 millibars, or 160 times weaker than that at the surface of the Earth. To have such thin air on our planet, it would be necessary to climb to an altitude of 25 miles (40 km). Under such a low pressure, water cannot exist in a liquid state. That meant saying goodbye to the beautiful dreams of oceans and rivers on Mars.

NASA launched two more spacecraft, *Mariner 6* and *Mariner 7*, toward Mars during the following "window of opportunity" in 1967. They flew by Mars at a distance of about 2,200 miles (3,500 km) on July 31 and August 5, respectively.

The two spacecraft took about 30 images each. These views were even closer than those sent by *Mariner 4*, but they were similarly disappointing: craters, craters and more craters. *Mariner 4* had photographed only about 1 percent of the Martian surface and *Mariner 6* and *Mariner 7* only 10 percent, with a resolution limited to 1.2 miles (2 km). Would the rest of the surface, still unexplored, reveal any surprises? Astronomers had little hope. However, they were wrong.

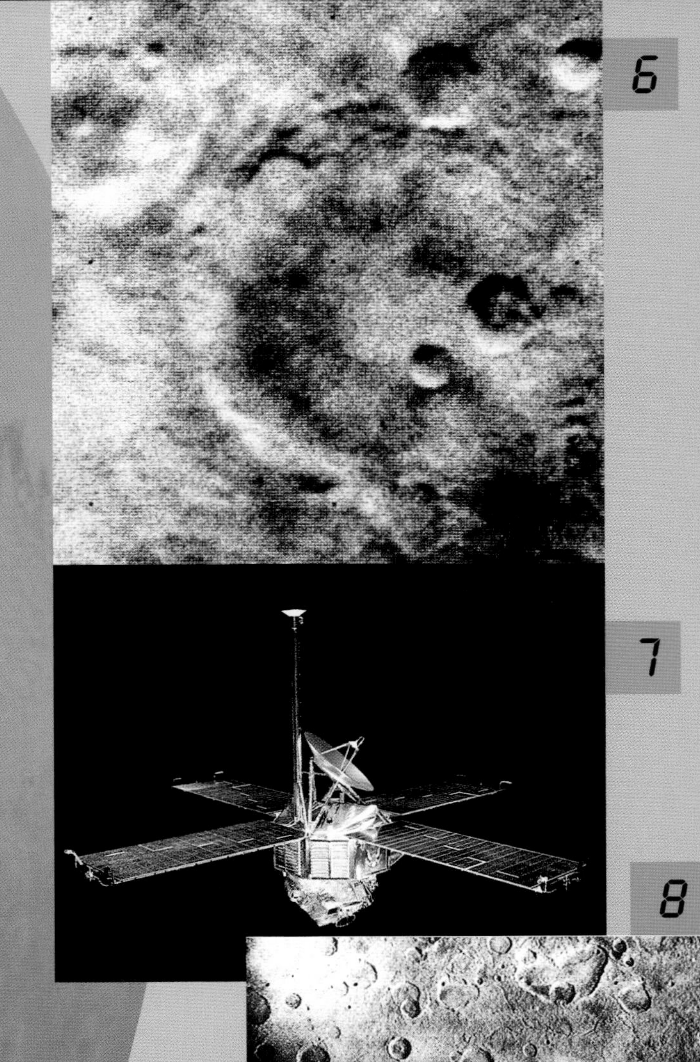

5

6

7

8

1 VENUS, CLOSER TO THE EARTH THAN MARS IS, WAS THE FIRST U.S. TARGET ...

2 ... WITH MARINER 2 *IN 1962.*

3 MARS 1 *WAS THE FIRST RUSSIAN SPACECRAFT SENT MARS IN 1962.*

4 ON THE WAY TO MARS: AN ATLAS ROCKET TAKES OFF IN 1964 ...

5 ... WITH THE MARINER 4 *SPACECRAFT ...*

6 ... WHICH PASSED NEAR MARS AND PHOTOGRAPHED THE PLANET IN JULY 1965.

7 MARINER 6 *AND ITS TWIN* MARINER 7 *FOLLOWED IN 1967 ...*

8 ... AND STILL SHOWED A DISCOURAGING LANDSCAPE OF CRATERS.

2 The Extraordinary Revelations of *Mariner 9*

Hope was revived in 1971. It became clear that the early *Mariner* spacecraft sent to Mars had directed their cameras at the least interesting regions. They had photographed the large plains of the Southern Hemisphere, unchanged for billions of years and still bearing the traces of the cataclysms that followed the solar system's formation.

With NASA's *Mariner 9* it became apparent that the equatorial and northern regions were very different in appearance. *Mariner 9* had time to explore the other regions of Mars. Rather than fly by very quickly, it went into orbit around Mars and photographed it for months, revolution after revolution.

The first weeks of observation showed little. Mars was almost completely hidden by clouds of dust kicked up by a gigantic storm. This phenomenon had already been observed from Earth by Gerard Kuiper in 1956. Storms often occur when Mars goes through its closest approach to the Sun during winter in the Southern Hemisphere. This shed light on one of the mysteries of the red planet's appearance: it's not covered by vegetation that changes color, but by dust that moves from region to region, darkening regions near the poles with a seasonal rhythm.

When the Martian sky finally cleared, *Mariner 9* found spectacular and unexpected formations: immense extinct volcanoes, dry riverbeds, gigantic canyons and more.

Mars is now a dusty desert – and very cold. During the northern winter, the temperature drops to –240°F (–150°C) at night. Near the equator, it very rarely reaches 85°F (30°C) during the day. But its past was very different. Several billion years ago, when life appeared on Earth, the red planet had a denser atmosphere and running water on the surface that carved riverbeds and canyons. What became of it? Did all of the water escape into space, or could some still be found underground, in a permanently frozen state reminiscent of Siberian permafrost? Did any life forms appear at that time? Are they still there in the subsoil, or in some particularly clement part of the red planet, where water might be present in liquid form during the summer days?

Mariner 9 revived hope in Lowell's dream – the idea of finding life on another planet relatively near the Earth. The "canals" are real, even if they aren't artificial structures; they are the beds of ancient rivers, reminiscent of the wadis of terrestrial deserts. And even if there aren't "intelligent" Martians, the discovery of microorganisms or their fossils would be enough to overturn our understanding of life in the universe.

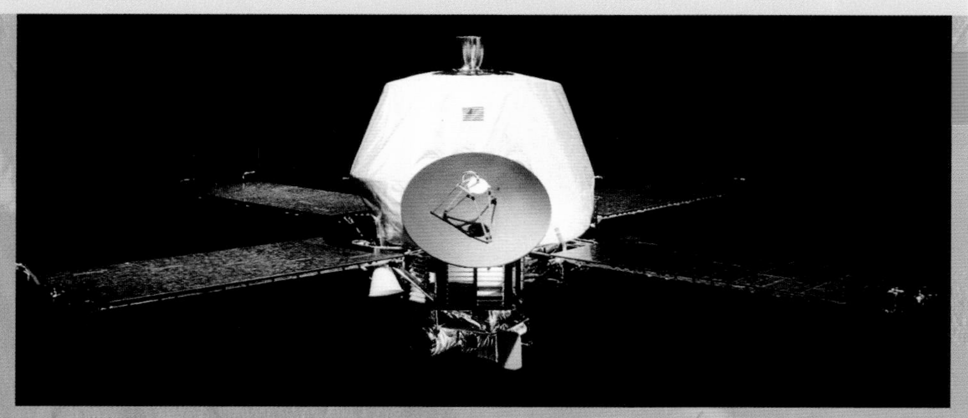

1

1 THE MARINER 9 SPACECRAFT WAS THE FIRST ARTIFICIAL SATELLITE OF MARS.

2 WHEN MARINER 9 DREW CLOSE TO MARS, ONLY ONE FORMATION WAS VISIBLE: THE GIANT VOLCANO, OLYMPUS MONS ...

3 ... BECAUSE THE PLANET WAS ENCASED IN AN ENORMOUS DUST STORM.

4 AFTER SEVERAL MONTHS, WHEN THE STORM SUBSIDED, THE PLANET'S RELIEF CAME INTO VIEW (HERE IN A SERIES OF VOLCANOES) ...

5 ... AND MARINER 9 WOULD TAKE IMAGES OF STUPEFYING QUALITY (HERE OLYMPUS MONS) ...

6 ... THAT WOULD REVEAL MARS' COMPLEX HISTORY.

7 VALLES MARINERIS, THE GIGANTIC COMPLEX OF CANYONS, SLOWLY COMES INTO VIEW THROUGH THE DUST CLOUDS.

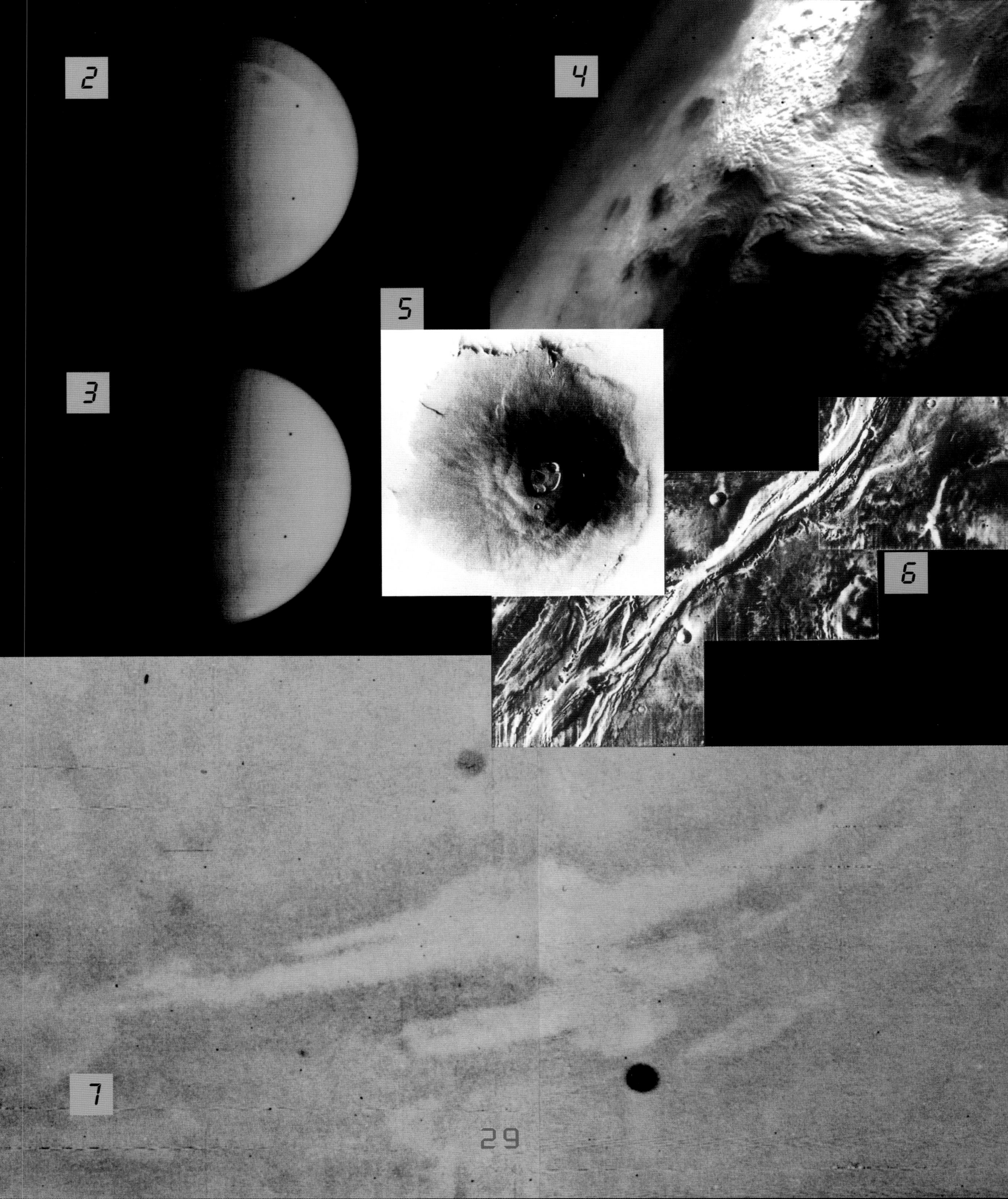
2
3
4
5
6
7

2 The Search for Life on Mars

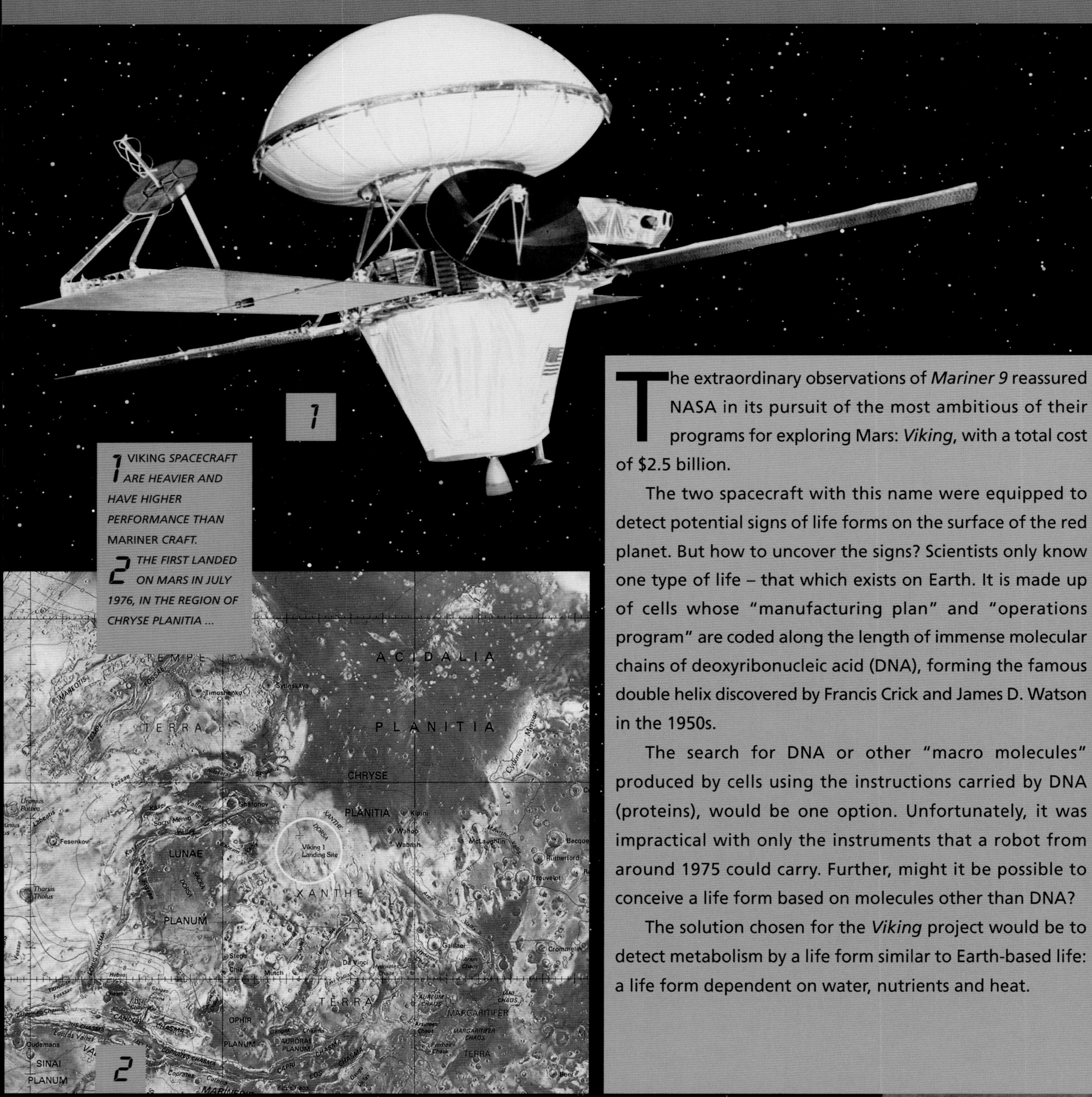

1 VIKING *SPACECRAFT ARE HEAVIER AND HAVE HIGHER PERFORMANCE THAN* MARINER *CRAFT.*

2 *THE FIRST LANDED ON MARS IN JULY 1976, IN THE REGION OF CHRYSE PLANITIA ...*

The extraordinary observations of *Mariner 9* reassured NASA in its pursuit of the most ambitious of their programs for exploring Mars: *Viking*, with a total cost of $2.5 billion.

The two spacecraft with this name were equipped to detect potential signs of life forms on the surface of the red planet. But how to uncover the signs? Scientists only know one type of life – that which exists on Earth. It is made up of cells whose "manufacturing plan" and "operations program" are coded along the length of immense molecular chains of deoxyribonucleic acid (DNA), forming the famous double helix discovered by Francis Crick and James D. Watson in the 1950s.

The search for DNA or other "macro molecules" produced by cells using the instructions carried by DNA (proteins), would be one option. Unfortunately, it was impractical with only the instruments that a robot from around 1975 could carry. Further, might it be possible to conceive a life form based on molecules other than DNA?

The solution chosen for the *Viking* project would be to detect metabolism by a life form similar to Earth-based life: a life form dependent on water, nutrients and heat.

3 ... USING RETROROCKETS TO LAND GENTLY ...

4 ... AFTER A PARACHUTE SLOWED ITS DESCENT.

5 THE VIKING LANDERS WERE LOADED WITH INSTRUMENTS INTENDED TO SEEK CHEMICAL MANIFESTATIONS OF METABOLISM FROM POSSIBLE MARTIAN MICROORGANISMS.

Viking 1 entered orbit around Mars on June 19, 1976. Its lander set down on a plain called Chryse Planitia a month later, on July 20. With its panoramic images, humanity discovered a reddish landscape, strewn with sharp, angular rocks. The sky, a pale salmon color, took on beautiful pink hues at sunset.

A remote-controlled arm gathered soil samples and placed them in incubators. The results of the analyses were ambiguous. For most scientists, they showed that the surface of Mars was completely sterile, but for others, the experiment might have indicated the growth of microorganisms.

The second *Viking* spacecraft landed on September 3, 1976, and arrived at the same uncertainty. Again, disappointment settled in. But how could it be any other way? Life on Mars could be very different from life on Earth and show up in surprising ways. It might also exist at only a few specific points on the surface, where water might occasionally be present in liquid form. It would be necessary to look elsewhere, in detail, and the *Viking* landers did not have that capability.

Despite its overly "biological" (in the terrestrial sense of the term) research program, the *Viking* experience greatly enriched our understanding of Mars. The two landers studied the rough Martian environment and sent 4,587 images showing their landing sites at all times of day and in all seasons. The two orbiters that circled Mars photographed it in detail. Their 51,539 images made it possible to map 97 percent of the Martian surface with a resolution better than 1,000 feet (300 m). They also confirmed the richness of the Martian relief with its volcanoes, craters, canyons, polar ice caps and so on.

Mars' past is enthralling and its present is undoubtedly thrilling.

1 *THE* VIKING *ROBOT ARM ...*

2 *... AND THE SCOOP USED TO SCRAPE THE SOIL AND GATHER SAMPLES.*

3 *A SUNSET ON MARS, PHOTOGRAPHED BY* VIKING.

MARS Pathfinder

Two decades after *Viking, Mars Pathfinder* ushered in a new way of exploring the red planet with smaller and cheaper spacecraft. On leaving the Earth, its mass was only 1,750 pounds (800 kg) and its development cost $265 million. *Mars Pathfinder* left the Earth on December 4, 1996, atop a Delta II rocket. Its journey lasted nearly 500 days, after which the probe entered Mars' atmosphere. Despite the action of parachutes and a retrorocket, its impact with the ground was hard. But *Mars Pathfinder* was protected by a unique system – 12 airbags, similar to those found in cars. The maneuver was successful, and on July 4, 1997, the United States placed a new operating spacecraft on the red planet.

1

1 ASSEMBLY OF THE MARS PATHFINDER SPACECRAFT IN A "WHITE ROOM."

2 TESTING THE 12 AIRBAGS THAT WERE TO SURROUND AND PROTECT THE CRAFT DURING ITS BUMPY ARRIVAL ON MARS.

3 MARS PATHFINDER WAS INSTALLED IN THE NOSE CONE OF THE DELTA II ROCKET THAT CARRIED IT TOWARDS THE RED PLANET.

2

3

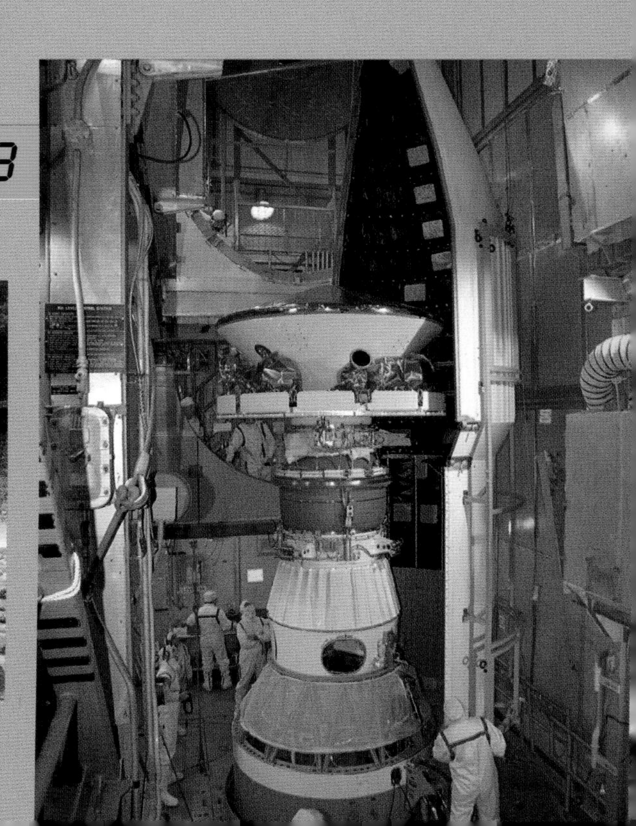

1 THE SMALL ROBOT SOJOURNER IS THE SIZE OF A RADIO-CONTROLLED CAR AND COVERED WITH A PANEL OF SOLAR CELLS.

2 A RAMP IS DEPLOYED TO ALLOW SOJOURNER TO DESCEND ONTO THE RED PLANET'S ROCKY SOIL.

III

1 MARS PATHFINDER ON THE RED PLANET, WITH ITS "PETALS" OPENED AND ITS ROVER ON THE MOVE.

2 THE LANDING SITE, AT THE MOUTH OF THE DRY RIVER ARES VALLIS IN THE CHRYSE PLANITIA REGION NOT FAR FROM THE VIKING 1 LANDING SITE.

1 SOJOURNER, THE SMALL MOBILE ROBOT, MAKES ITS WAY IN FRONT OF ONE OF THE DEFLATED AIRBAGS. ON THE HORIZON, THE FORMATION CALLED NORTH TWIN PEAK, ABOUT 2,800 FEET (860 M) AWAY, RISES UP ABOUT 100 FEET (30 M).

2 THE SUN, ABOUT TWICE AS SMALL AS SEEN FROM THE EARTH, HOVERS OVER THIS PANORAMA PHOTOGRAPHED BY MARS PATHFINDER; LOOKING TO THE SOUTHWEST, THE POINTED FORMATION CALLED SOUTH TWIN PEAK IS LESS THAN A MILE (1 KM) AWAY. THE SMALL ROBOT SOJOURNER STUDIES A ROCK ON THE EDGE OF A ROCKY ZONE, CALLED ROCK GARDEN. SOME OF THESE ROCKS HAVE ROUNDED EDGES, A LITTLE LIKE PEBBLES, WHICH COULD MEAN THAT THEY HAD BEEN TRANSPORTED BY SIGNIFICANT FLOWS OF WATER.

1 A CLOSE-UP OF YOGI, SHOWING A DUSTY ROCK WITH A SURPRISING BLUISH RED COLOR ON THE RIGHT.

2 IT TOOK ABOUT 10 HOURS FOR SOJOURNER TO ANALYZE THE COMPOSITION OF THE ROCK.

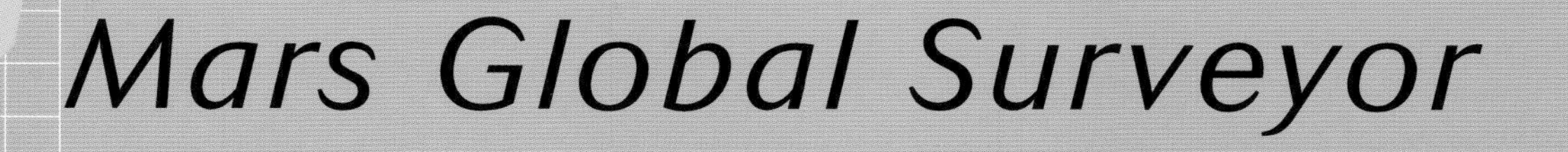

Mars Global Surveyor

Mars Global Surveyor successfully left Cape Canaveral on November 7, 1996. *Mars Pathfinder*'s turn came the same year on December 4. These spacecraft belong to a new series of craft that are smaller and less expensive than *Viking* and make as much use as possible of new technologies. Although it left first, *Mars Global Surveyor* arrived at its destination a little after *Mars Pathfinder* – on September 12, 1997. It started a long and delicate braking maneuver, using the Martian atmosphere to move itself into a low orbit around the planet.

From 1999 on, it photographed specific parts of the Martian surface with unprecedented resolution (about 3 feet/1 m). Its observations showed a planet even stranger than had been thought. Many formations seemed to show the presence of water, certainly in a frozen state, just a little below the surface. These formations actually have no equivalent on Earth, and all the hypotheses put forth to explain them have still not been proven.

1

1 MARS GLOBAL SURVEYOR *POINTS THE TELESCOPIC LENS OF ITS CAMERA TOWARDS THE RED PLANET ...*

2 *... AND PROVIDES SCIENTISTS WITH IMAGES OF EXCEPTIONAL QUALITY THAT NOTABLY REVEAL THE COMPLEXITY OF THE MARTIAN CANYONS.*

2

The "Great Martian Vampire"

After the remarkable exploits of *Viking*, it was 21 years before another spacecraft landed on the red planet. One cause of this long gap in the exploration of Mars was NASA's lack of interest. After the apparently negative results of *Viking*'s analyses, disappointed by the "sad" Moon discovered by the Apollo astronauts and discouraged by the mysterious nature of Mars, NASA set its objectives for exploration on other objects, the distant planets.

But another factor was also present – a succession of failures. Some people found humor in claiming that a "great Martian vampire" was behind the failures. The Soviets counted 13 failures in total between 1960 and 1973. The most frustrating was the *Mars 3* landing on December 2, 1971, four-and-a-half years before the *Viking* craft. The transmissions by *Mars 3* from the surface of the red planet only lasted 20 seconds! Another attempt in 1973, this time with four heavy spacecraft from the same class as *Viking*, failed completely. The Soviets would abandon Mars for 15 years.

When they set out on the path again in 1988, in collaboration with several countries including France, there was a magnificent program for studying Mars, especially its satellite Phobos. The two *Phobos* spacecraft failed their missions. Another eight years went by. The Soviet Union collapsed, but Russia picked up with another very international project – *Mars 96*, which finished its trajectory in the Pacific Ocean. After 16 tries, all ending in failure, the Soviet/Russian Mars programs stopped, perhaps forever.

NASA wasn't spared either. In 1992, their timid return to Mars, with the orbiting probe *Mars Observer*, failed three days before the craft was to go into orbit. The "great Martian vampire" was keeping a careful watch. It would act again at the beginning of 1998 and 1999, when NASA lost two new spacecraft: *Mars Polar Orbiter and Mars Polar Lander*. Fortunately, before these two spectacular failures, the remarkable successes of both *Mars Pathfinder* and *Mars Global Surveyor* in 1997 rekindled interest in studying the planet Mars.

1 THE SPECTACULAR LAUNCH OF THE MARS POLAR LANDER SPACECRAFT ON JANUARY 3, 1999.

2 THE SOVIET MARS 3 SPACECRAFT TOUCHED DOWN ON MARS ON DECEMBER 2, 1971 ...

3 ... BUT ITS SIGNALS STOPPED AFTER 20 SECONDS.

4 SOVIET/RUSSIAN FAILURES: 16.

5 NASA WOULD HAVE BETTER LUCK IN 2003 WITH TWO NEW LANDERS...

6 ... FOLLOWING THE TOTAL LOSS OF MARS POLAR LANDER IN JANUARY 2000?

7 THE SOVIETS HAD NO BETTER LUCK WITH THEIR STUDY OF PHOBOS, PRINCIPAL SATELLITE OF MARS, IN 1988.

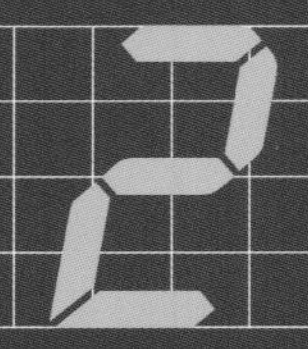

Mars on Earth

1

2

Pieces of Mars – on Earth? This surprising discovery was established in 1968: three meteorites, found near the towns of Shergotty, India, Nakhla, Egypt, and Chassigny, France, are believed to have come from the red planet.

On Mars as on Earth, cataclysmic events sometimes occur. On Earth, the impact of an asteroid or comet caused the extinction of the dinosaurs 65 million years ago. The impact of a celestial object, even of modest dimensions, can blast pieces of the surface loose and throw them into space.

That is the hypothesis proposed for SNC meteorites (named after the initials of the three towns near which they were discovered). After having spent millions of years around the Sun, these pieces of Mars met up with the Earth and fell to our planet's surface. Currently there are 30 of them. It is difficult to be entirely sure that these meteorites came from Mars. However, an early indicator is given by their composition: a basaltic lava that cooled slowly; this phenomenon is only possible within a large planet, and Mars is the most probable source. The most reliable proof has come from comparison with the analyses of the Martian atmosphere performed by the *Viking* landers: bubbles containing air from their original planet were discovered in the SNC meteorites, and the composition of that air turned out to be close to the atmosphere of Mars.

One of the more recent SNC meteorites was discovered in Antarctica, where "meteor hunters" find the best conditions for their searches. In this area, meteorites stand out visibly from the frozen surface of the Antarctic deserts, after violent winds expose them at the surface following thousands of years of burial.

The size and shape of a small potato, meteorite ALH84001 was collected in 1984, but its membership in the SNC group was not recognized until 1994. Torn from Mars 16 million years ago, it fell to Earth 13,000 years ago. A group of scientists, under the direction of David McKay of NASA, conducted studies whose results were published on August 7, 1996. Those results raised both enthusiasm among the public and controversy in scientific circles: ALH84001 revealed, in particular, traces of Martian microfossils, shaped like worms, but a hundred times smaller than the smallest bacteria known on our planet. Other parts, like the presence of minuscule globules of calcium carbonate, could also be of biological origin.

Did McKay actually discover traces of past life on Mars? At the beginning of the 21st century, the question remains unanswered.

1 METEORITE HUNTERS AT WORK ON THE ANTARCTIC ICE ...

2 ... WHERE PIECES OF ROCK FALLEN FROM THE SKY ARE EASILY VISIBLE DESPITE THEIR SMALL SIZE.

3 THE METEORITES ARE UNCOVERED BY EROSION OF THE ICE THAT PRESERVED THEM FOR THOUSANDS OF YEARS.

4 TERRESTRIAL ROCKS 3.5 BILLION YEARS OLD RETAIN TRACES OF STICK-SHAPED MICROBACTERIA.

5 COLONIES OF MICROBACTERIA ARE ALSO FOSSILIZED IN STILL OLDER ROCKS (3.8 BILLION YEARS OLD).

6 SCIENTISTS WONDER IF MARTIAN FOSSILS OF "NANOBACTERIA" ...

7 ... MIGHT HAVE BEEN FOUND IN METEORITE ALH84001.

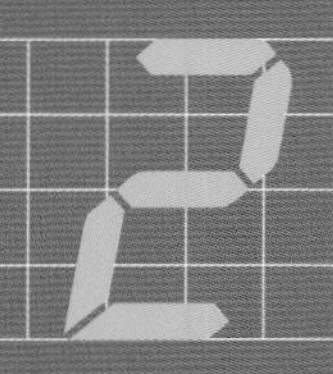

New Approaches to Mars

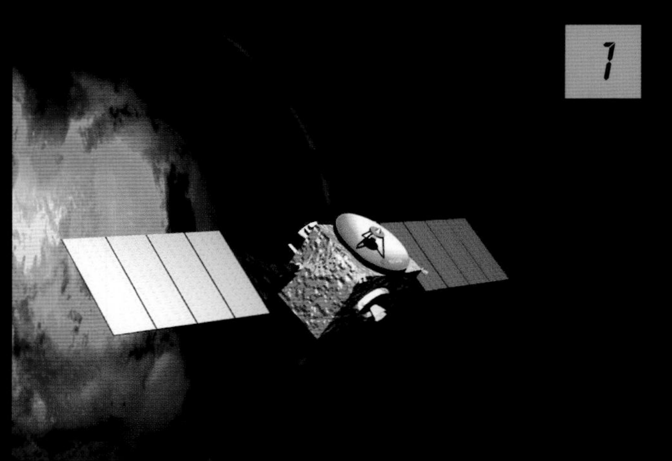

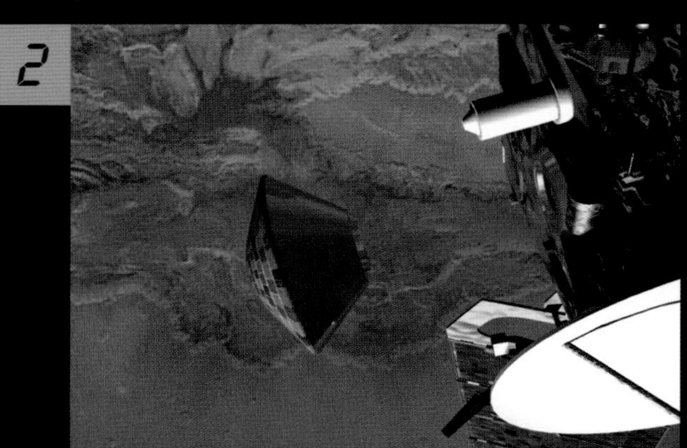

1 CHRISTMAS 2003: THE EUROPEAN SPACECRAFT MARS EXPRESS *APPROACHES THE RED PLANET ...*

2 ... TOWARDS WHICH IT LAUNCHES A CAPSULE WITH A SMALL LANDER.

Even if European scientists have been associated with many experiments involving the exploration of Mars, the United States and Russia (strengthened by the heritage of the Soviet Union) maintained their monopoly throughout the 20th century.

The beginning of the third millennium was marked by Europe's entry into space exploration of the red planet. In June 2003, the *Mars Express* spacecraft of the European Space Agency (ESA) took off.

Mars Express weighed one ton upon departure from the Earth. The orbiter carried seven instruments that would investigate the Martian atmosphere, surface and even subsurface. It would also launch a lander six days before it entered orbit around Mars. The lander, built by the British, was named *Beagle 2* in honor of the ship Charles Darwin sailed. It took full advantage of new electronic and mechanical technologies that allowed the construction of a 65-pound (30 kg) laboratory equipped with a stereoscopic camera, a microscope and a soil analysis system, using x-rays and a minuscule robotic "mole."

ESA's spacecraft joined two NASA spacecraft: *Mars Global Surveyor*, which continued the photographic mission that started in 1997, and *Mars Odyssey 2001* (named in honor of the famous film directed by Stanley Kubrick), in orbit since October 2001.

Mars Express reached the red planet in December 2003—a wonderful Christmas present for ESA astronomers. Soon, however, there was bitter disappointment. After *Beagle 2*'s rough ride down to the surface of Mars, ESA scientists were unable to raise a signal from it. After repeated efforts to make contact had failed, the Beagle team declared the lander to be lost in February 2004.

By then, two NASA exploration rovers had captured the world's attention: *Spirit*, which was launched June 10, 2003 and landed on Mars January 3, 2004, and *Opportunity*, which was launched July 7, 2003, and landed on Mars January 24, 2004. Spirit explored the terrain of the Gustav Crater, while *Opportunity* surveyed the Meridani Planum. Both rovers sent breathtaking panoramic views back to Earth. This was followed by even more exciting news: *Opportunity's* analysis of minerals showed that a salty sea had once existed at its landing site.

Meanwhile, ESA's *Mars Express* was circling the planet on a path that brought it 155 miles (250 km) above Mars every seven hours. It obtained stunning high-resolution views with a German stereoscopic camera and a French cartographic camera. Its onboard spectrometer confirmed the presence of small amounts of methane in the Martian atmosphere. This was an exciting development since, on Earth, methane is among other things a byproduct of biological processes.

The United States next plans to put a satellite intended to take very high-resolution images (*Mars Reconnaissance Orbiter*) into orbit around Mars in 2005, and then to place a long-duration rover—a mobile scientific laboratory—on Mars in 2009.

Cooperation between the United States and Europe will continue beyond 2010 with an experiment of great interest: the arrival on Earth of Martian soil samples.

3 *NASA LANDED* SPIRIT *AND* OPPORTUNITY *IN JANUARY 2004.*

4 *... BUT* BEAGLE 2 *WAS LOST.*

3 From Astronomy to Astronautics

Until the end of the first half of the 20th century, the Moon and planets were the private domains of astronomers. First with the naked eye, then with lenses and telescopes of increasing power, they unceasingly probed the mysteries of distant bodies across the immensity of space.

Astronomy is an observational science – the object of study is mostly inaccessible. It inspires the imaginations of artists and poets, as shown by the lyricism of Camille Flammarion's texts or the unbridled imagination of Percival Lowell. But exploration has a very different nature, which associates research with understanding of risk, adventure and, not to mention, the quest for riches. Astronomy and exploration joined in 1957, giving birth to astronautics – "navigation between the stars." Space was no longer just an object of study; it became an area where robots and then people could go on the path to the Moon and planets.

During the first half-century of the Space Age, astronauts only visited the Moon, but all the planets, except for Pluto, were approached by spacecraft. Some spacecraft became artificial planetary satellites or landed on planet surfaces in our solar system.

With robots acting as intermediaries, astronomers became explorers, and the astronomers now encountered risks and undertook long journeys reminiscent of the epoch of the great geographic exploration of Earth.

To move from astronomy to astronautics, scientists needed to find a means to cross the barrier of the Earth's gravity and go off into space. It was the great Russian pioneer Konstantin Tsiolkovski who imagined the means of moving in space – the rocket, which enables propulsion in a vacuum, entry into orbit around the Earth and travel in the solar system.

The rocket has existed in Asia since A.D. 1000. Later, in the 15th century, it was used for military purposes in Europe; the English, in particular, used it in battles with Napoleon. But before Tsiolkovski, no one had realized that this instrument of death (and also of entertainment – it is used in fireworks) could work outside the atmosphere and serve as a means of propulsion for spacecraft. This use is just a simple consequence of Newton's fundamental laws of dynamics, and could have been discovered in the 18th century.

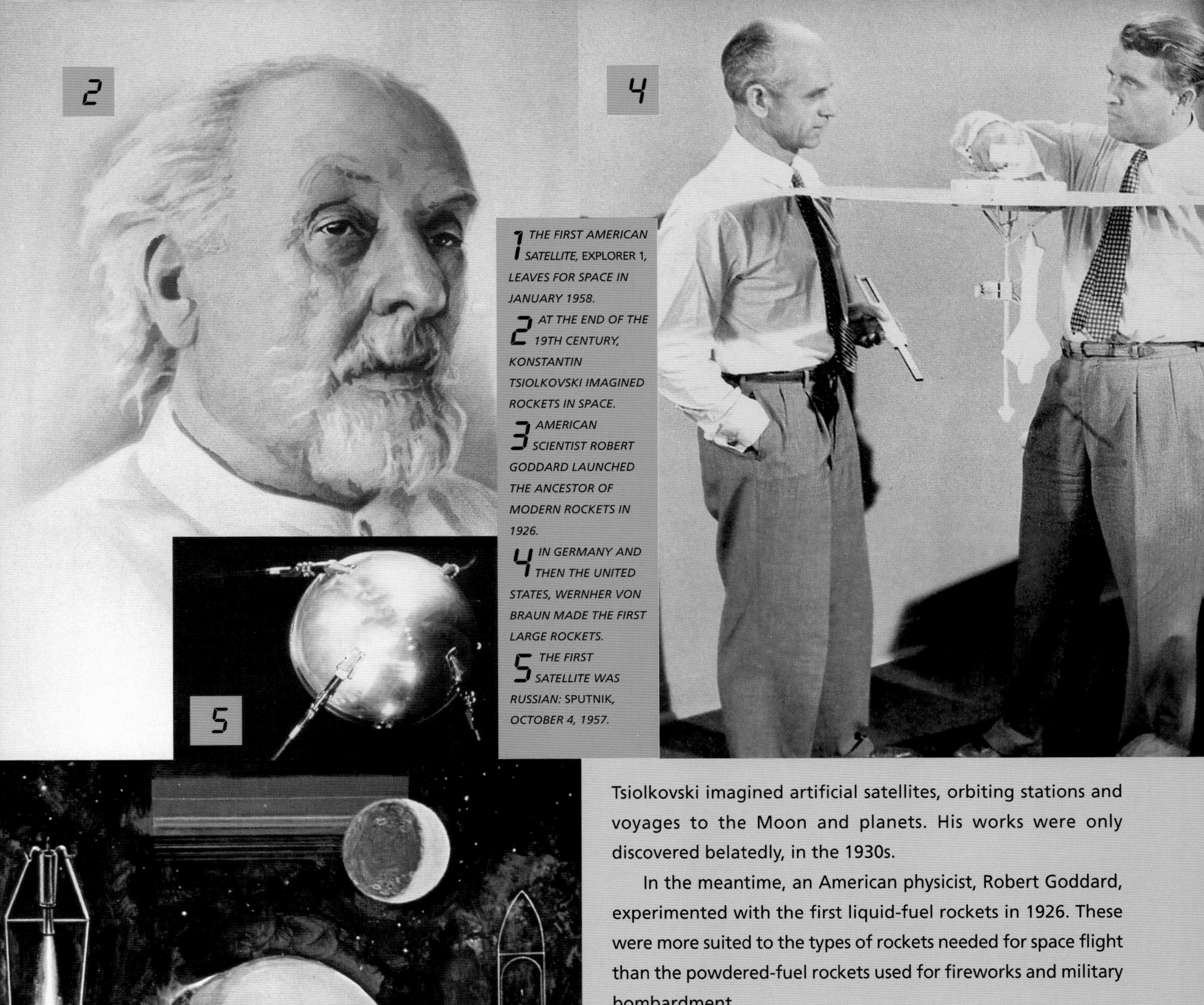

1 THE FIRST AMERICAN SATELLITE, EXPLORER 1, LEAVES FOR SPACE IN JANUARY 1958.

2 AT THE END OF THE 19TH CENTURY, KONSTANTIN TSIOLKOVSKI IMAGINED ROCKETS IN SPACE.

3 AMERICAN SCIENTIST ROBERT GODDARD LAUNCHED THE ANCESTOR OF MODERN ROCKETS IN 1926.

4 IN GERMANY AND THEN THE UNITED STATES, WERNHER VON BRAUN MADE THE FIRST LARGE ROCKETS.

5 THE FIRST SATELLITE WAS RUSSIAN: SPUTNIK, OCTOBER 4, 1957.

Tsiolkovski imagined artificial satellites, orbiting stations and voyages to the Moon and planets. His works were only discovered belatedly, in the 1930s.

In the meantime, an American physicist, Robert Goddard, experimented with the first liquid-fuel rockets in 1926. These were more suited to the types of rockets needed for space flight than the powdered-fuel rockets used for fireworks and military bombardment.

Then events accelerated. In Germany, Wernher von Braun made the first long-distance rocket, the V-2, which could already reach an altitude of 50 miles (80 km) and a speed of 3,000 mph (5,000 km/h). After 1945, the V-2 inspired the Cold War combatants (the United States and Soviet Union) to build "intercontinental" missiles. These missiles were not only capable of crossing 5,000 miles (8,000 km) in half an hour, but also of launching artificial satellites.

On October 4, 1957, the Soviet craft *Sputnik* circled the Earth. The U.S. in turn launched *Explorer 1* into orbit on January 31, 1958. Humanity had entered the Space Age.

First Steps in Space

1 THE FIRST HUMAN WENT INTO SPACE EARLY IN THE MORNING OF APRIL 12, 1961.

2 YURI GAGARIN BECAME THE HERO OF THE SOVIET PEOPLE.

3 JOHN GLENN, THE FIRST AMERICAN TO ORBIT THE EARTH, HAD HIS TURN IN FEBRUARY 1962.

4 PRESIDENT JOHN F. KENNEDY WITH GLENN NEXT TO THE MERCURY CAPSULE..

From the first years of the conquest of space, astronautics split into two distinct branches – on the one hand, missions with robots and, on the other, human voyages in space. The first were launched to study and to put into use practical applications such as the observation of the Earth and telecommunications, while the second extended the great geographic explorations and prepared for the expansion of humanity into the universe.

In the 20th century, the exploration of Mars and the other planets of the solar system was the domain of robots, from *Mariner 2* to *Mars Global Surveyor*. But when the small automatic craft *Mariner 2* was launched towards Venus in 1962, humans had already traveled around the Earth.

Though latent human interest in space had already been seen at the end of the 19th century with the popular success of Percival Lowell's and Camille Flammarion's ideas about Martian life, it suddenly became evident with the launch of *Sputnik*. Throughout the world, *Sputnik*'s journey was greeted with enthusiasm, and the two superpowers of the time, the United States and the Soviet Union, understood that the people of the world would judge their respective powers by the measure of their exploits in space.

1

2

The exploration of space became the most fantastic race in human history. All the objectives were pursued at once: launching artificial satellites around the Earth, space exploration of the Moon and planets by interplanetary craft and, especially, sending human beings into space.

The "Christopher Columbus of the 20th century," Yuri Gagarin, was a young Russian, 26 years old. He flew from the Baikonur Cosmodrome in Central Asia early in the morning of April 12, 1961. He traveled in the first spacecraft in history, *Vostok*, created, like its launch rocket, by an engineering genius, Sergei Korolev, the "father" of Soviet astronautics. Korolev and Wernher von Braun were the two men who made the largest contribution to the movement of humanity into space.

Korolev did not hesitate to take risks to make Gagarin the first man in space. He would be compensated for his efforts: *Vostok*'s journey around the Earth was an immense triumph, the greatest that the Soviet Union would ever know. The trip was short: only one orbit about the Earth, lasting a total of 108 minutes. But when Gagarin landed by parachute on the great Russian plain near Saratov, the world had changed. Humanity had started on the path into space.

The United States did not delay in duplicating Gagarin's feat. John Glenn completed three orbits around the Earth in his *Mercury* space capsule in February 1962. But for several years, the Soviets would lead the dance in the great cosmic ballet: the first day-long voyage by Gherman Titov in August 1961; the first joint flight of two spacecraft, with the "brothers in space" Andrian Nikolayev and Pavel Popovich in August 1962; the first voyage of a woman in space: Valentina Tereshkova in June 1963; the first flight of a spacecraft with three cosmonauts – *Voskhod 1* in October 1964; and, finally, in March 1965, Aleksei Leonov, the first cosmonaut to leave his spacecraft.

3 Meeting Above the Earth

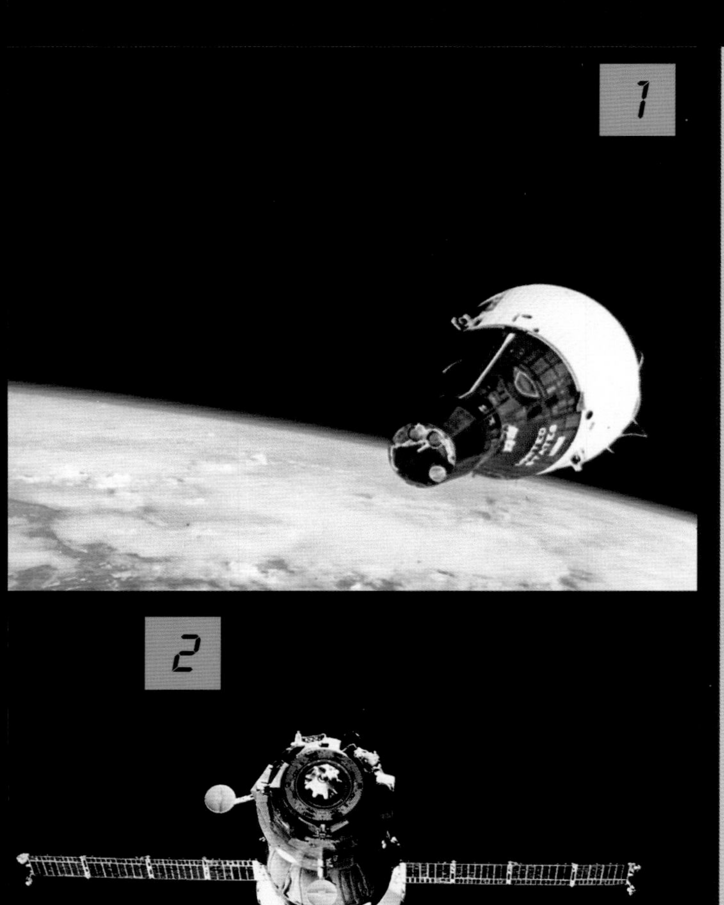

The first spacecraft, *Vostok* from the Soviet Union and *Mercury* from the United States, were simple capsules placed in orbit around the Earth by their booster rockets. They were not capable of modifying their trajectory to actually "navigate" in space. They could only brake their movement with a retrorocket, to leave orbit, reenter the atmosphere, be slowed by friction with the air and touch down on land or sea using a parachute.

The cosmonauts (the term used by the Russians) and the astronauts (the term the Americans prefer) were simple passengers. At most, they could orient their capsule in space and trigger ignition of the retrorocket. To perform more complex missions, like trips to the Moon or the planets, or construction of orbiting stations, the capsules had to give way to higher performance vessels.

The latter were capable of changing their trajectory, maneuvering in space to leave their initial orbit and reach a near or distant object (for example, other space equipment) or to go to the Moon or Mars. Cosmonauts and astronauts became "pilots" and carried out "piloted space missions."

The first piloted spacecraft in history was the American *Gemini* spacecraft, so named because it carried two astronauts. Its conical shape recalled the *Mercury* capsule, but its capacities were much greater. *Gemini* had small rocket motors, which it could use to modify its orbit around the Earth and accomplish the fundamental maneuver of astronautics – a meeting in space,

1 *THE FIRST MEETING IN SPACE:* GEMINI 6 *JOINED* GEMINI 7 *IN DECEMBER 1965.*

2 *THE RUSSIAN* SOYUZ *PROGRAM, COMPETING WITH* GEMINI, *DID NOT BEGIN UNTIL 1967.*

3 *LAUNCHED BY* TITAN *ROCKETS, THE* GEMINI *CAPSULES FOLLOWED EACH OTHER AROUND THE EARTH IN 1965 AND 1966.*

which involves joining another vehicle in space and, if necessary, docking with it. This operation would make it possible, for example, to reach a base installed in orbit around the Earth, assemble a large spacecraft from smaller elements or even abandon one craft that has completed its mission to take a new one – a little like leaving a car at a parking garage before taking a train.

The *Gemini* flights started in March 1965, right after the Russian cosmonaut Aleksei Leonov exited his vessel in space.

4

5

6

4 A MINUSCULE COMPARTMENT FOR TWO ASTRONAUTS.

5 THE FIRST AMERICAN SPACEWALK, IN MAY 1965, TWO MONTHS AFTER THAT OF RUSSIAN ALEKSEI LEONOV.

6 THE NASA CAPSULES TOUCH DOWN IN WATER ON THEIR RETURN.

Extravehicular activity in space is another fundamental operation: Cosmonauts and astronauts in spacesuits assemble equipment in space and, if necessary, fix it. In May 1965, the Americans repeated Leonov's exploit with Edward White's spacewalk. But the big objective was the meeting in orbit. It was achieved in December 1965 with a meeting of Gemini 6 and *Gemini 7*. The first docking followed in 1966 between *Gemini 8* and an automated vehicle, *Agena*. With the Gemini missions, the United States largely took the lead in the space race. The Soviets prepared a comparable vessel: *Soyuz*, but it was not until 1967.

When the last *Gemini* spacecraft returned to Earth in December 1966, NASA had mastered all of the techniques necessary to move to the following phase in the conquest of space: sending a human being to the Moon.

3 Apollo on the Moon

1 THE GIANT SATURN 5 ROCKET SENDS THE FIRST AMERICAN ASTRONAUTS TO THE MOON.

2 BUZZ ALDRIN DESCENDS TO JOIN NEIL ARMSTRONG ON THE SURFACE OF THE MOON, JULY 20, 1969.

3 NEIL ARMSTRONG: THE FIRST PERSON ON THE MOON.

4 STARTING WITH THE APOLLO 15 MISSION IN 1971, THE ASTRONAUTS MOVE AROUND WITH A LUNAR ROVER.

5 THE HEROES RETURN.

A few weeks after of Yuri Gagarin's triumphal flight on May 25, 1961, American President John F. Kennedy made a historic declaration: The United States would send a man to the Moon before the end of 1969.

The statement was audacious. At the time, the American space agency, NASA, had not yet sent an astronaut into orbit around the Earth, and its rockets were much less powerful than those available to the Soviets. But compared to the progress needed to go to the Moon, the Russian advantage in space was slim. That's where President Kennedy's challenge made sense: Choose such an ambitious objective that only American technology could achieve it.

The Gemini program's successes in 1965 and 1966 had already revealed the potential of this technology. All the same, it was only a beginning. To fully carry out the Apollo program for conquest of the Moon, astronautics would have to change in scale.

To go to the Moon, the Americans needed a giant rocket 10 times more powerful than Gagarin's *Vostok* launcher. They built the *Saturn 5*. Measuring 363 feet (111 m) high and weighing 3,000 tons on launch, it was built under the direction of Wernher von Braun. This great German pioneer, a naturalized American after World War II, found the opportunity to play an essential role in the realization of his great dream: the human conquest of the solar system. The first *Saturn 5* took off from Cape Canaveral in May 1967. Its flight was a great success. Unfortunately, it came right after a tragedy. In January 1967, the first crew of an *Apollo* spacecraft, training for an experimental mission around the Earth, died during a fire in the capsule. That catastrophe would delay the Apollo program's progress for a year and a half but would not stop the United States in their mission to reach the Moon.

Events unfolded quickly starting in the fall of 1968. The first flight of an *Apollo* spacecraft around the Earth, *Apollo 7* (the others had been automated missions), came in October. The flight of *Apollo 8* took a crew of three astronauts into orbit around the Moon in December. For the first time, astronauts passed behind the Moon and could not see the Earth.

The dress rehearsal was performed around the Earth in March 1969, using both parts of the *Apollo* spacecraft: the Command Service Modules, and a Lunar Module. On board the first, the three astronauts would complete the round-trip flight between the Earth and an orbit around the Moon. The Lunar Module itself was going to leave that orbit with two astronauts to land on the Moon before returning to the Command Service Modules using a meeting in lunar orbit.

A first complete *Apollo* spacecraft was sent to the Moon in May 1969, but its Lunar Module was content to skim past the Moon 9 miles (15 km) away.

The first lunar landing was reserved for the astronauts of the *Apollo 11* spacecraft's Lunar Module. On July 20, 1969, Neil Armstrong and Buzz Aldrin walked on the Moon. Humanity had accomplished a great step on the road to the stars. In total, the Apollo program included six successful missions to the Moon, the last of which took place in December 1972. The crew of the *Apollo 17* were the last astronauts to dig the lunar soil in the 20th century.

5

3 What to Do After the Moon: The Long Pause

No one had imagined that the conquest of the Moon would happen so quickly. After the extraordinary success of the Apollo program, a question arose: What to do next? NASA had a plan largely inspired by Wernher von Braun: Set up on the Moon, build a large space station to orbit Earth and leave for Mars very quickly – a mythical objective. The United States government decided against it. The country had made a gigantic effort to send astronauts to the Moon – it now had other priorities. Mars could wait.

NASA had to watch its ambitions decline. With its greatly restricted means, it concentrated on building a space airplane, the shuttle, capable of making round trips between the Earth and nearby space. Intended to later service an "orbiting station" – a complex installed to remain in space – it would greet astronauts who could live and work there for months, even years.

For their part, the Soviets had also tried to send cosmonauts to the Moon with their giant *N1* rocket, the maneuverable *Soyuz* craft and a lunar lander closely resembling the *Apollo* Lunar Module. But it failed because the *N1* could not be made to work properly. Also, following the example of the Americans, they needed to set new, more modest objectives. In 1969, they gave priority to the construction of orbiting stations.

After their great lunar programs, the winner and loser of the race to the Moon returned close to the Earth. The champions of human expansion into space passed away: Sergei Korolev in 1966 and Wernher von Braun 10 years later. So began a long pause in human exploration of the solar system. But the 21st century would prove to be favorable for a new departure in this formidable adventure, with the experience of long-duration space flights, advanced technologies and renewed ambitions.

1 THE RUSSIANS ALSO DREAMED OF THE MOON ...

2 ... AND CONSTRUCTED THE GIANT N1 ROCKET SHAPED LIKE A PYRAMID ...

3 ... WHICH WAS ASSEMBLED IN A GIANT FACTORY AT BAIKONUR ...

4 ... AND THEN TRANSPORTED HORIZONTALLY TO THE LAUNCHPAD.

5 BUT ALL THE N1 LAUNCHES WERE FAILURES.

II
Stations and Shuttles

After the conquest of the Moon, a new adventure began: orbiting stations. They involved "houses in space," circling a few hundred miles above the Earth, and serving both as housing and a research laboratory. The big novelty was the length of the missions: comfortably installed in vast quarters, with all the conveniences (toilets, showers), even individual rooms, astronauts and cosmonauts went to work for weeks and then months in orbit.

Of course, compared with a lunar landing, endlessly circling the Earth is not as exciting. But to prepare to go farther, to make an expedition to Mars or the other planets possible, learning how to live in space for long intervals is necessary. As well, orbiting space stations had made important studies – involving the Sun, the universe, the Earth as seen from space, plus matter and life in the absence of gravity – possible.

The first orbiting station came from the Soviet Union: Salyut 1, in 1971. But its first crew died during the return to Earth. NASA had the first success: the Skylab station, where crews stayed one, two and then three months. The Soviets then remained alone on the path with their Salyut stations (up to number seven) and then the large space station Mir, which operated from 1986 to 2001 (the station became Russian in 1991). Among their crew, some had stayed over a year.

1

2

1 SKYLAB, ABOVE THE EARTH IN 1973–74.

2 THE SOVIET/RUSSIAN STATION MIR, WITH ITS MANY MODULES.

1 The Soviets also constructed a shuttle, Buran, launched for the first time in 1988 by the giant Energia rocket. But the dissolution of the Soviet Union in 1991 led to the abandonment of the program.

2 Beginning in 1972, NASA dedicated all its efforts to the construction of a space shuttle that could carry 10 astronauts and 20 tons of freight into space, while serving as an embryonic orbiting station. It flew for the first time in 1981.

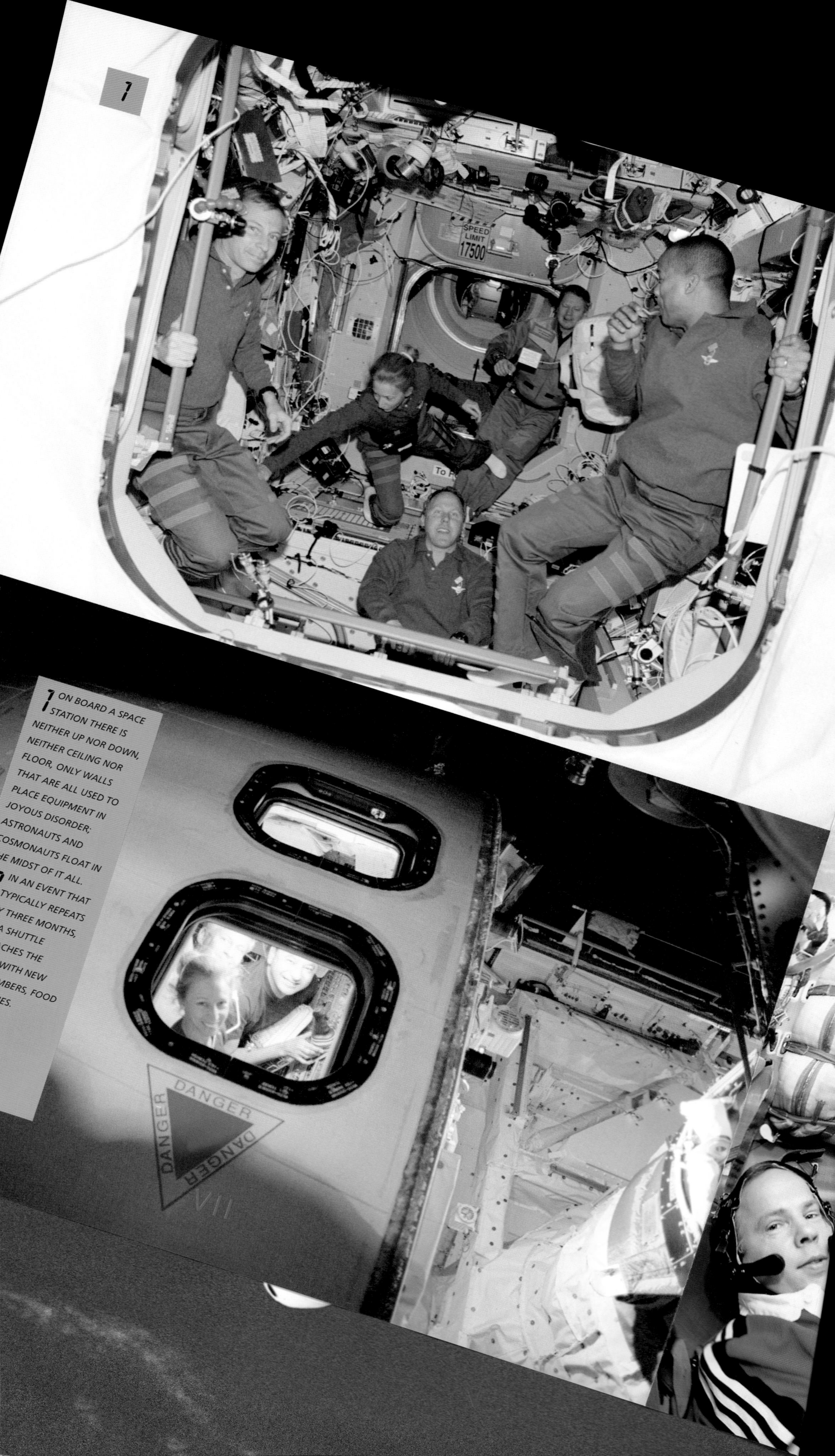

1 ON BOARD A SPACE STATION THERE IS NEITHER UP NOR DOWN, NEITHER CEILING NOR FLOOR, ONLY WALLS THAT ARE ALL USED TO PLACE EQUIPMENT IN JOYOUS DISORDER; ASTRONAUTS AND COSMONAUTS FLOAT IN HE MIDST OF IT ALL. 2 IN AN EVENT THAT TYPICALLY REPEATS Y THREE MONTHS, A SHUTTLE CHES THE WITH NEW MBERS, FOOD ES.

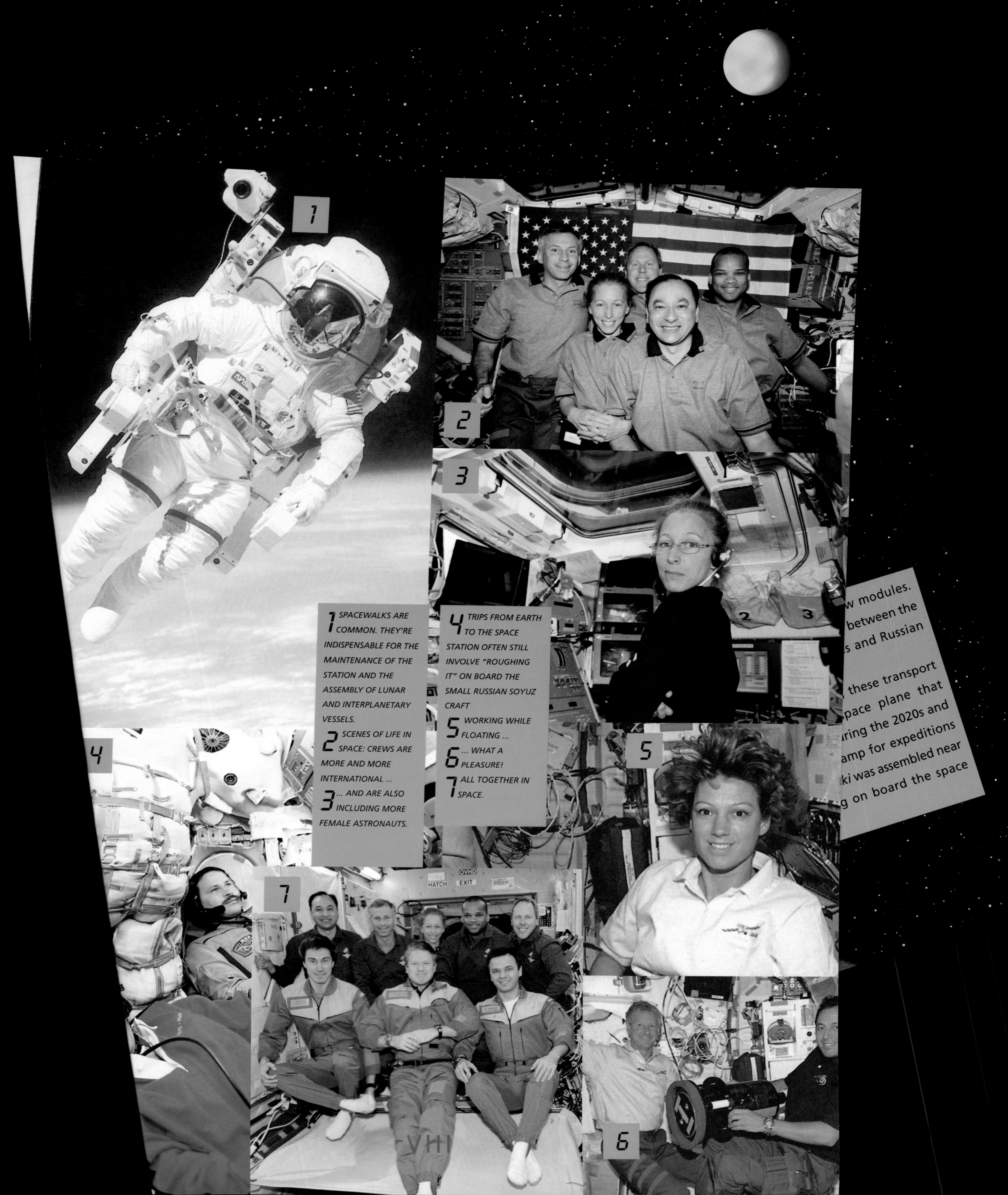

1 SPACEWALKS ARE COMMON. THEY'RE INDISPENSABLE FOR THE MAINTENANCE OF THE STATION AND THE ASSEMBLY OF LUNAR AND INTERPLANETARY VESSELS.
2 SCENES OF LIFE IN SPACE: CREWS ARE MORE AND MORE INTERNATIONAL ...
3 ... AND ARE ALSO INCLUDING MORE FEMALE ASTRONAUTS.
4 TRIPS FROM EARTH TO THE SPACE STATION OFTEN STILL INVOLVE "ROUGHING IT" ON BOARD THE SMALL RUSSIAN SOYUZ CRAFT
5 WORKING WHILE FLOATING ...
6 ... WHAT A PLEASURE!
7 ALL TOGETHER IN SPACE.

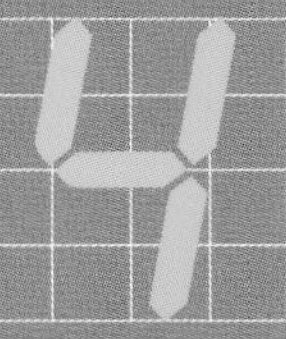

The Decision to Go to Mars

1 IN 1961, PRESIDENT KENNEDY SAID HE WOULD SEND ASTRONAUTS TO THE MOON ...

The major decisions about space exploration are political. In 1961, President John F. Kennedy solemnly announced the launch of the Apollo program before the United States Congress. In 1972, Richard Nixon decided to begin construction of a space shuttle. Ronald Reagan chose a new objective in 1984 – the construction of a large space station – and invited his country's allies to join a project that would go on to become the International Space Station (ISS).

The first announcement of an expedition to Mars came in July 1989 from President George Bush. At the Johnson Space Center in Houston, Texas, on the 20th anniversary of the first walk on the Moon, he committed to his Space Exploration Initiative (SEI), which aimed to send astronauts to Mars by 2019 – 50 years after man had set foot on the Moon. But his initiative was a moot issue. With the technologies available at the end of the 20th century, a voyage to Mars would take too long and be too expensive. In January 2004, President George W. Bush reaffirmed the American intention to return to the Moon and after that, to go to Mars.

The technical progress achieved during the first two decades of the 21st century made a voyage to Mars feasible. But the operation of the ISS in particular, which started in 2000, was to open the path to the internationalization of larger space activities. In 2007, and on the 50th anniversary of *Sputnik*, the World Space Agency (WSA) coordinated the programs of the national (Japan, Canada, Brazil and India) and regional (greater Europe, with which Russia is associated) agencies. China initially pursued an independent program, but it didn't take long for them to join the rest of the world in what would become a global undertaking, destined to build the future of humanity in space.

The available technologies were then suitable for the project, and the international context was favorable. But why give priority to a voyage to Mars instead of a base on the Moon or a large robotic exploration program of the distant planets? The passion for Mars had been awakened by *Mars Pathfinder*, *Mars Odyssey 2001* and all the robotic missions launched to Mars at the beginning of the 21st century. But the truly decisive factor was the return of Martian soil samples to the Earth in 2014. Fossil microorganisms, identified beyond doubt, confirmed the earlier studies conducted with Martian meteorites.

The time to hesitate was over. In October 2017, during an international summit, the leading members of the WSA decided to send astronauts to Mars before 2027. The mission was a little delayed, but now, in 2033, the *Tsiolkovski* is on its way with a crew of two women and five men.

2 ... AND PRESIDENT GEORGE BUSH COMMITTED TO THE SPACE EXPLORATION INITIATIVE (SEI) IN 1989.

4 Why People?

The international decision to send a crew to the red planet reopened an old debate, nearly as old as the conquest of space: Should people or robots be sent? Wouldn't robots be capable of replacing people in the exploration of the cosmos, and also be more economical and perhaps more effective?

The dominant environment in space, like that on the surface of the Moon or Mars, is hostile: empty, cold, filled with radiation and so on. It's necessary to create a protective shell around an astronaut with a spacecraft, housing, vehicle and spacesuit. It's also necessary to meet their physiological needs, including air, water and food. And, unlike a robot, astronauts cannot be miniaturized. For all these reasons, an inhabited spacecraft has a mass 100 to 1,000 times greater than an automated spacecraft. The robot that brought samples from the surface of Mars back to Earth had a mass of only 6 tons when it left for Mars. The *Tsiolkovski*, with its crew of seven astronauts, has a mass of 300 tons, and was preceded to the red planet by two cargo ships that were even heavier.

The presence of people is expensive – very expensive. But is it justified? Compared to the *Viking* and *Pathfinder* landers, or to the vehicles that rolled on the Martian surface during the first decade of the 21st century, immense progress has been made in robotics.

Moore's Law, stated in 1966, predicted the capacities of electronic circuits would double every 18 months. Developing with the cadence of that celebrated observation, the robots of 2029 are 100 million times more "intelligent" than those of 2000. They also take advantage of the progress in "micromechanics," which enables them to have supple parts, traverse obstacles, climb, dig, and so on. The robots of 2029 are marvels. They have even changed life on Earth. After occupying factories and then taking over all transportation, robots have also entered our homes, where they perform all of the domestic chores.

However, despite these fantastic abilities, they still don't have, and undoubtedly will never have, the essential capacity of a human being: creative intelligence. Its nature remains a mystery, linked to the extraordinary complexity of the mind and the human body, and is the result of billions of years of evolution of life on Earth.

Art and science are the exclusive domains of human beings. In laboratories, computers and robots are fantastic tools for serving researchers. But the spark of intelligence that once leapt from the mind of Isaac Newton or Albert Einstein and changed the world exceeds the understanding and possibilities of even the most sophisticated electronic systems.

A good example can be found in paleontology, where researchers must quickly recognize that the minuscule object they see is not a rock, but part of a fossil. Only the combination of the exceptional capacities of the eye and hand with the paleontologist's creative intelligence allow the discovery of the fossil. Further, the search for life on Mars presents more difficulties than those found in paleontology. We don't know what we might be looking for, and it's necessary to have the idea, the creative spark to see that what might resemble, for example, a pebble, is in fact a fossil or living being. With so many unknowns, a person is indispensable.

1

1 THE SMALL ROBOT SOJOURNER FROM MARS PATHFINDER IN 1997 ...

2 ... HAD MUCH MORE CAPABLE SUCCESSORS ...

3 ... BUT HUMAN INTELLIGENCE IS IRREPLACEABLE IN THE SEARCH FOR FOSSILS, BOTH ON MARS AND EARTH; PALEONTOLOGISTS' FIELDWORK CANNOT BE AUTOMATED.

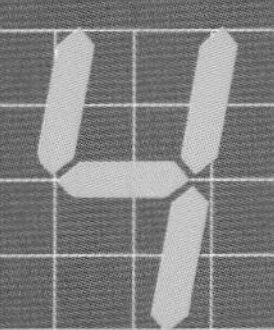

Von Braun's Dream

1 WERNHER VON BRAUN IN THE EARLY 1950S WITH A MODEL OF THE GIANT ROCKET HE DREAMED OF BUILDING.

One man, more than any other, played a decisive role in human exploration of space and in the conquest of Mars: Wernher von Braun. Creator of the V-2 rocket during World War II, he was picked up by the United States at the end of the conflict, along with his principal lieutenants, in an operation called "Paperclip" that was worthy of a spy novel. The young German engineer's past – he was not yet 40 when he crossed the Atlantic – had many shadowy areas, linked to the fabrication of the V-2 in a concentration camp at Dora. But the Americans glossed over these facts and put this man's technical genius into their service. However, he was not given responsibilities equal to his capacities. He simply was required to improve his small V-2, which became the Redstone, while others made the large intercontinental missiles that were to transport the first astronauts into space.

Thus marginalized at the beginning of the 1950s, von Braun had time to dream. He imagined a vast plan for the human conquest of space at a time when robots only existed in science fiction, computers occupied entire buildings and only astronauts could pilot a spacecraft. The first step in his plan would be the installation of an immense, wheel-shaped orbital station 1,050 miles (1,700 km) above the Earth, with a crew of 80 people. Trips between the Earth and this station would be made with an enormous rocket weighing 14,000 tons that would send a giant plane into space. The space plane would be capable of carrying 30 astronauts and would glide back to Earth once the mission was done. Von Braun's wheel-shaped station inspired the one in Stanley Kubrick's magnificent film *2001: A Space Odyssey*. It would be the point of departure for a great expedition to the Moon, with several gigantic, spherical craft carrying dozens of astronauts. However, the major objective of von Braun's plan was to be able to reach Mars – an armada of craft would fly to the red planet, and gliders would land on Mars using skis.

Von Braun was not content to dream. He communicated his plan and his enthusiasm to the public at large through a series of articles published in 1953 in *Collier's* (a magazine with a very large circulation), with marvelous illustrations by the painter Chesley Bonestell, and then in television broadcasts produced by Walt Disney.

Von Braun became a star, and his plan was adopted by NASA, which hoped to work on it after the success of the Apollo program. Of course the shuttles, space stations and craft were smaller – economic reality imposes limits – but the program was the same. The "von Braun paradigm" has remained the foundation of space programs with human crews in the 21st century.

2 THE RANGE OF SPACECRAFT IMAGINED BY VON BRAUN: A WHEEL-SHAPED SPACE STATION, AND LUNAR AND MARTIAN SHIPS.

3 LARGE-SCALE LUNAR EXPEDITIONS ...

4 ... FOLLOWED BY A GREAT MISSION TO MARS.

5 WERNHER VON BRAUN WITH A MODEL OF HIS LUNAR SHIP.

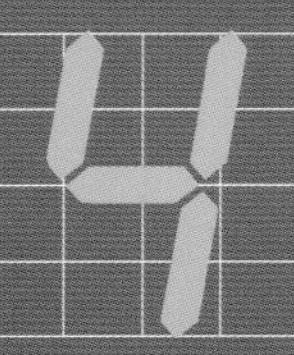

4 NASA's Abandoned Plans

In the limelight of late 1969 following the successful landing of Neil Armstrong and Buzz Aldrin on the Moon, NASA proposed an ambitious "post-Apollo program," which incorporated all the elements of Wernher von Braun's plan. It anticipated the construction of a space shuttle intended to replace all of the rockets, the installation of a base on the Moon, the construction of a large space station and a trip to Mars by astronauts — all in less than two decades.

It's true that NASA succeeded in sending astronauts to the Moon in just eight years. Its program had certainly been technically feasible ... but at a price of hundreds of billions of dollars! The United States Congress and the American government reduced the NASA budget, so they needed to get by with much more modest projects – the space shuttle and the International Space Station – to the great disappointment of everyone excited about space.

The United States, however, did not totally renounce von Braun's plan. In 1986, President Ronald Reagan created the National Commission on Space; it made some proposals on the future of America in space the following year. The program was still the same: set up on the Moon and leave for Mars. For this commission, it was clear that the future of humanity was the occupation of the solar system, and that the United States had to construct, in space, the infrastructure allowing for the colonization of the Moon and planets. It had to play the role in space that the railroad and then the interstate highways had played in America.

The Commission included a space station near the Earth, a lunar base and stations placed in highly strategic points or orbits for transportation in the solar system, for example, on trajectories around the Sun regularly passing near the Earth and then Mars – an Earth-Mars "Beltway" of sorts. A crew would simply have to take a small shuttle to reach such a transport station when it passed near the Earth, live there the length of the voyage, and then go down to Mars on another small shuttle. The idea is brilliant. It will undoubtedly be made in the second half of the 21st century, but in 1987 it was far too advanced for its time and for the means that the United States was ready to dedicate to space.

At the beginning of the 1990s, strengthened by the conclusions of the National Commission on Space and then by the support of President George Bush in 1989 with his Space Exploration Initiative, NASA again prepared a program of human exploration of the red planet, but it fell victim to the same pitfalls: too long, too expensive. The truth is that neither public opinion nor technology was ready for the great Martian voyage. It would be necessary to wait until 2017 to reconsider the question, and 2033 for the project to come into being with the *Tsiolkovski*.

1 IN THE 20TH CENTURY, NASA DID NOT SUCCEED IN CARRYING OUT THE LARGE PROJECTS FROM THE 1980S AND 1990S, LIKE THE RETURN TO THE MOON AND THE ESTABLISHMENT OF A SMALL BASE THERE ...

2 ... WHICH LED ENTHUSIASTS LIKE ROBERT ZUBRIN, PRESIDENT OF THE MARS SOCIETY, TO PROMOTE THE IDEAS OF A RAPID VOYAGE TO MARS – BUT IN VAIN. IT WOULD BE NECESSARY TO WAIT DECADES FOR THE TSIOLKOVSKI *TO FINALLY SET OUT TO MARS.*

Like Being on Mars ... in the Far North

Still brushed off by the American authorities, NASA's great Martian plans of the 20th century left the entire community of space fanatics frustrated; they had dreamed of immediate voyages to the red planet. The great driver behind this "Martian movement" was an aerospace industry engineer, Robert Zubrin, who imagined a mission to Mars that was very simplified compared to the NASA projects. It could be done in less than 10 years for $10 billion – a sum much less than that spent for the Apollo program. The spirit of this mission was like that of the conquerors of the New World, who had set sail in the 17th century, without guarantee of a return, and onboard uncomfortable ships. The Mars Direct mission was to use the giant Russian rocket, *Energia*, which had flown only twice in 1987 and 1988, to send the elements of a Martian ship, into space. To reduce the weight that was to be launched toward Mars, Zubrin proposed touching down on the red planet in a ship without the propellant necessary for the return to the Earth – it would be manufactured on Mars from carbon dioxide gas in the atmosphere.

The concept of using extraterrestrial resources was an excellent one, and was adopted for the Tsiolkovski mission. The technology has progressed a lot now in 2033 and has been tested on Mars by robots. However, the gamble in the 1990s was much too audacious. Further, the costs and intervals proposed for Mars Direct were much too optimistic. The red planet would have to wait to welcome its first visitors until a program could be set up that was both ambitious and realistic.

Another of Robert Zubrin's ideas, carried out by the Mars Society that he created and ran, had greater success: preparing for future Mars missions by setting up prototype Martian bases in extreme environmental areas, like polar regions and tropical deserts. On the one hand, there was cold and wind, on the other dryness and heat. Of course, no terrestrial environment really corresponds to conditions on Mars, with its combination of very thin air, complete absence of water and temperatures much lower than the most frozen points on Earth, but it involved subjecting men and women to life and work in an isolated base in a hostile environment, as a base on Mars would be. It was also possible to run the base within a closed circuit, as the situation would be on Mars, and to use spacesuits and sealed vehicles for outside activities. The first of these prototype Martian bases on Earth was established in Canada on Devon Island and has operated since 2000.

1

1 "MARSONAUTS" ... ON EARTH! ...

2 ... IN AN EXPERIMENTAL HABITAT CONSTRUCTED ON DEVON ISLAND, WHERE THE ISOLATION, DRYNESS AND COLD CARRY A SLIGHT RESEMBLANCE TO THE LIVING CONDITIONS FOR FUTURE EXPLORERS OF MARS

3 EXCURSIONS WITH PROTOTYPE SPACESUITS FOR MARS.

4 ROBERT ZUBRIN, THE PRESIDENT OF THE MARS SOCIETY, PROMOTED THE OPERATION.

5 A MODULE RESEMBLING FUTURE ELEMENTS OF A BASE ON MARS.

6 A SMALL VEHICLE FOR EXPLORATION.

7 HELMETS OFFER GOOD VISIBILITY.

2

3

4

5

6

7

4 The Delayed Revolution of Space Transport

Why, in view of choosing not to go to the Moon and Mars, did astronauts have to be content for nearly half a century with endlessly circling the Earth? The principal reason was economic: space transport remained very expensive – too expensive.

Going into space had necessarily been expensive because a rocket could be used only one time. What would the price of a flight from Paris to New York have been if an airplane, like an Airbus, were tossed on the scrap heap after each crossing of the Atlantic? In 2001, a rocket like the *Ariane 5* cost as much as an Airbus – more than $120 million. It was able to send a useful load of 10 tons towards the International Space Station (ISS), so the price of freight transport between the Earth and the station was about $6,000 per pound. Since a ship destined for Mars has a mass of around 300 tons before leaving the Earth, the simple transport of these elements to the space station would cost, under those conditions, more than $3.6 billion.

Beyond any doubt, the sum was excessive. NASA attacked this problem in the 1970s with its space shuttle program by using a "reusable" launch vehicle, which could shuttle between the Earth and near space. NASA had hoped to reduce the price of space transport by a factor of 50, but that objective proved to be illusory. Some parts of the shuttle – the space plane called orbiter – was effectively reusable, but the cost of preparing it for use remained prohibitive: about $480 million per mission. With this system, which transported a maximum of 10 people, the price of sending astronauts to the ISS was at least $48 million. Considering such costs, trips to the Moon and Mars were inconceivable, and even using

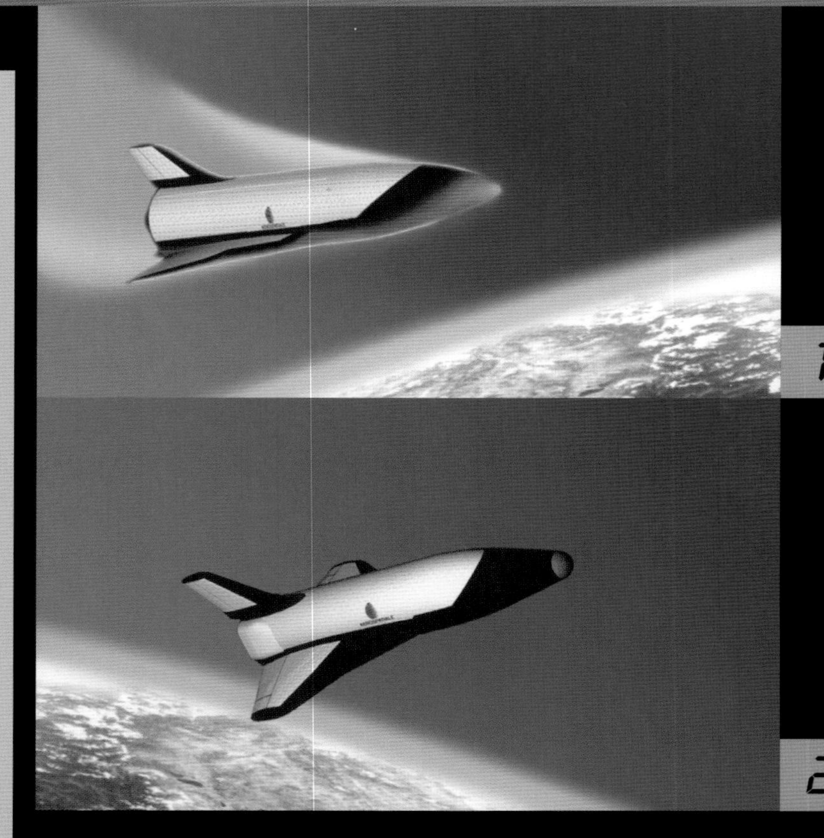

the ISS remained too expensive.

The problem was not unsolvable, but it had, in any case, been underestimated. In the early 21st century, going into space remained an arduous undertaking at the edge of existing technologies. Reducing the cost of access to space required a major technological effort, with new thrusters and new materials. Two steps were still necessary. In 2012, a new shuttle, which was entirely reusable, and produced in cooperation by the United States, Europe, Russia and Japan, was placed into service. With two airplane-shaped stages still powered by rocket engines, it could service the space station for just 20 percent of the cost of the original shuttle.

1 REUSABLE LAUNCHERS OF THE FUTURE WERE BEING STUDIED IN EUROPE …

2 … BY THE EADS (EUROPEAN AVIATION DEFENSE AND SPACE) COMPANY.

3 THE ENGINEERS DREAMT OF A SPACE PLANE WITH A SINGLE STAGE THAT REACHED NEARBY SPACE …

4 … AND RETURNED TO LAND ON EARTH, THEN WAS READY TO IMMEDIATELY LEAVE AGAIN.

3

4

Commercial versions derived from this "two-stage shuttle" appeared around 2020. They replaced conventional rockets like *Ariane 5* and allowed considerable development in the practical application of satellites.

The second stage came in 2025, in time for the Tsiolkovski mission: an entirely reusable launcher with a single stage could finally accomplish round trips between Earth and space in a manner similar to a cargo plane. It used a rocket motor for takeoff and during the final phase of its flight into the cosmic vacuum, and a new "hypersonic combustion stratoreactor" type thruster, burning oxygen from the air while traversing the atmosphere. The costs were reduced by a factor of over 20 compared to the year 2000: $300 per pound of freight and $1 million per astronaut. It was still too expensive for true "space tourism," but the route for travel to Mars was opened, especially because the new techniques for interplanetary transport were also improved during the first decades of the 21st century.

4 A Stop on the Moon

Is a stopover on the Moon necessary to get to Mars? Unlike the space station in orbit 280 miles (450 km) above the Earth, the Moon is not a required stop-off point for the conquest of the solar system. Astronauts visited it first for a simple reason: It was nearby. With the rockets available in the 20th century, only three or four days were needed to get to the Moon, whereas a voyage from the Earth to Mars took six to nine months depending upon the relative positions of the two planets.

Even so, going to the Moon is not easy. The Earth's natural satellite is a large body, which strongly attracts approaching objects. (This attraction is the origin of the tides on Earth.) A probe launched from the Earth toward the Moon would crash into the lunar soil at a speed of 5,600 mph (9,000 km/h) if it did not make use of a braking system. There is no atmosphere to make use of aerodynamic friction, and a retrorocket must be used in difficult conditions; if the vehicle's velocity does not reach exactly zero at the surface, it's a catastrophe. For these reasons, going to the Moon is just as costly in energy use as going to Mars, but it's the

1

2

3

distance and the duration of the round trip that make the voyage to the red planet difficult.

Returning to the Moon decades after the end of the Apollo program was only justified by the interest it generated in its own right. Actually, it makes a remarkable platform for setting up large instruments to study the universe. It's possible to set up networks of telescopes connected together by laser beams so that their images can be combined – astronomers call it "optical interferometry" – which should be able to allow observation of planets around other stars. It's also possible to set up immense radio telescopes on its hidden face, which is sheltered from the Earth's "electromagnetic pollution" (all radio, television, radar and other emissions). The Moon is also ideal for experimenting with technologies developed for the Mars mission: nuclear reactors, variable power plasma thrusters, habitation modules, vehicles and so on.

Therefore, the plan for exploring Mars (established in 2017) called for the installation of a small scientific base on the Moon in 2025, where astronauts would train for the big voyage. Later, in the 2030s, it was to become a real forwarding station of humanity in the cosmos.

1 *A CYLINDRICAL HABITAT MANIPULATED BY A SMALL CRANE ...*

2 *... WAS USED TO BUILD A FORWARDING POST ON THE MOON, SERVED BY VEHICLES ...*

3 *... LIKE THOSE THAT WOULD LATER BE USED FOR THE EXPLORATION OF MARS.*

4 *ESTABLISHING A SMALL BASE ON THE MOON HAD BEEN STUDIED BY THE U.S. ...*

5 *... AND THE RUSSIANS, STARTING IN THE 1970S.*

4

5

4 The Assembly of the *Tsiolkovski*

Despite the immense technical progress achieved during the first decades of the 21st century, the ships required for the conquest of Mars are gigantic: several hundreds of tons each, nearly as heavy as the large International Space Station (ISS). In total, over 1,000 tons are needed for the three ships: two cargo ships, which transported the equipment needed to establish a large base for living and research on Mars, and the *Tsiolkovski*, which carries the crew – the first seven astronauts to voyage to Mars.

The three ships did not leave at the same time and did not follow the same route. The two cargo ships – named *Oberth 1* and *Oberth 2* in memory of the great German rocketry pioneer Hermann Oberth – were placed on a trajectory that was slow and economical, like all automated craft that have taken the path to Mars since 1962. Travel time was about seven months, but that hardly mattered; unlike the astronauts, the equipment being transported did not suffer from the long voyage, and it arrived well before the astronauts did.

Oberth 1 left the Earth in January 2029 to reach the red planet in July and land its payload on the surface in September. *Oberth 2* left April 12, 2031, marking the 70th anniversary of Yuri Gagarin's voyage. The equipment that it transported will be ready for use on Mars before the end of the year.

Next, the assembly of the *Tsiolkovski* began near the ISS. There was no real possibility of launching such a vehicle directly from the Earth. Like the two cargo ships before it and the Earth-Moon shuttles, it was built in orbit around the Earth, from smaller components launched by giant rockets descended from the old, Russian launcher *Energia*. The idea of using the new, very economical aerospace planes – which now provide transport to the space station and launch automatic satellites to provide multiple services for the inhabitants of Earth – was abandoned. These planes have a limited useful load: only 20 tons. It would take dozens of launches to transport the pieces and propellant for a single ship destined for Mars.

This procedure had been used for the ISS between 1998 and 2006. About 50 launches of American shuttles and Russian Protons had been needed. But the operation had been so long, complex and risky that no one wanted to repeat it. With *Energia*, three to five missions were sufficient for each Mars ship. The final assembly and filling of tanks was accomplished by robots and a few astronauts based in the ISS.

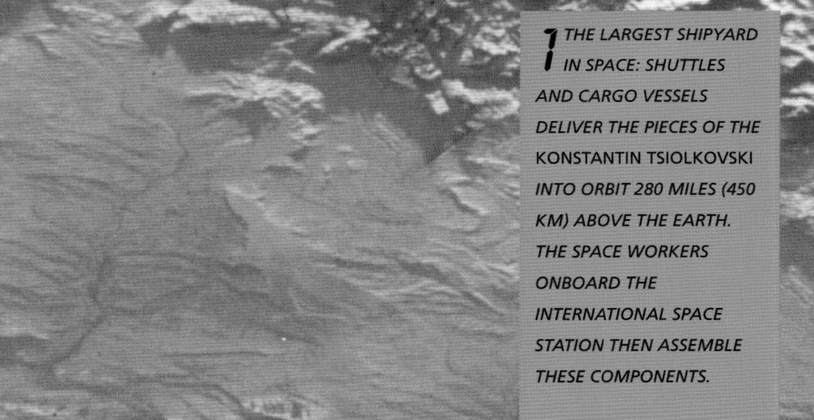

1 THE LARGEST SHIPYARD IN SPACE: SHUTTLES AND CARGO VESSELS DELIVER THE PIECES OF THE KONSTANTIN TSIOLKOVSKI *INTO ORBIT 280 MILES (450 KM) ABOVE THE EARTH. THE SPACE WORKERS ONBOARD THE INTERNATIONAL SPACE STATION THEN ASSEMBLE THESE COMPONENTS.*

1
TSIOLKOVSKI
UNITED NATIONS

5 Who Will Be the Crew?

Who will have the honor of being the first person to set foot on Mars? Even when the crew of seven and their two alternates was selected in July 2029 – 12 years after the decision to send people to the red planet and 60 years after the Apollo triumph – nobody knew yet. Mars' first ambassador from planet Earth will be chosen by drawing names in the final moments, just after landing on the surface of Mars.

The crew was chosen with both political and technical issues in mind. This is an international mission, created under the umbrella of a world agency, but the main effort is being accomplished by a few countries. The United States is represented by two astronauts, among them the onboard commander. Europe, whose space effort grew at the beginning of the 21st century, also has two crew members. Russia, always very advanced in matters of space technology, has one representative among the astronauts, just like Japan and Canada, which are historic partners in the International Space Station.

Why seven astronauts instead of three or 12? Such a choice is always arbitrary, but the seven-person crew was considered the minimum to bring together all the technical and scientific competencies needed for the mission's success. A doctor was needed and there was no discussion about it – a trip of such a length, with no possibility of return in case of an emergency, had a significant chance of being affected by illnesses or accidents. Therefore, the doctor, Eduardo Duarte, an American citizen but a native of Brazil, has a fundamental role. During the training phase, his biological expertise will allow him to also become an expert in astrobiology, with the knowledge necessary to study extraterrestrial life.

The mission's scientific leader, Michel Morey from France, is an astrophysicist specialized in the study of the surface of Mars.

The other astronauts were selected based on their in-depth knowledge of the systems supplied by their country. The advanced propulsion system used for the voyage between the Earth and Mars was developed in the United States and holds no secrets for the mission commander, John Sturgett. The nuclear energy sources were built in Russia and are the specialty of the second-in-command, Natasha Titova. The lander was designed in Europe, with significant German contribution. Their expert is an engineer, Otto Kruger. The astronauts' habitat on Mars was developed in Canada, and their astronaut is architect Jeannette Noordung. Finally, Japan built the base camp structures that will be placed on Phobos, the last step before the red planet. Their representative, Nagatomo Itochu, is ready to install the equipment that will manufacture the propellant necessary for the return to Earth from existing materials on Phobos,. Seven astronauts for the greatest voyage of all time – an excellent team.

III
The Ship for a Great Voyage

Ready for Departure

2029 to 2033: Four years of training were required for the group of seven astronauts and their two alternates, completed in the three large world centers for preparing space travelers: Star City near Moscow, NASA's Johnson Space Center in Houston, Texas, and the European Astronaut Centre in Cologne, Germany. The expedition does not pose any specific physiological problems from a medical standpoint. The trips between Earth and Mars, in both directions, are short compared to the long-duration flights completed on the Russian Salyut and Mir space stations in the 1980s and 1990s. These trips are less than five months in duration – short compared to the 437 days physician Valery Poliakov spent on board Mir, which is still the world record for weightless flight. Why carry out longer weightless missions, now that we have the technology allowing quick trips to Mars and, later, to Jupiter?

The other phases of the mission, which would have required physical conditioning with 20th-century technologies, are now easy to endure. Passengers fly between the Earth and the International Space Station (ISS) in new single-stage shuttles that are as comfortable as airplanes. The acceleration toward Mars, braking on arrival around the red planet, the descent to the surface, and the acceleration and deceleration for the return to Earth all take place very gradually. In the training centers, the emphasis is therefore on science and technology rather than sport. The only exception is training in swimming pools, with spacesuits on, to prepare for possible spacewalks during the voyage. This training for outside work is continued with real spacewalks at the ISS and in-depth training on the Moon in handling spacesuits and Martian vehicles.

The program is exhausting, but in the spring of 2033, the crew is ready for its historic journey.

1 A SINGLE STAGE SHUTTLE TAKES THE CREW OF THE TSIOLKOVSKI INTO SPACE ...

2 ... AFTER THEIR TRAINING ON EARTH, WHICH INCLUDED WORKING IN SWIMMING POOLS TO EXPERIENCE WEIGHTLESSNESS.

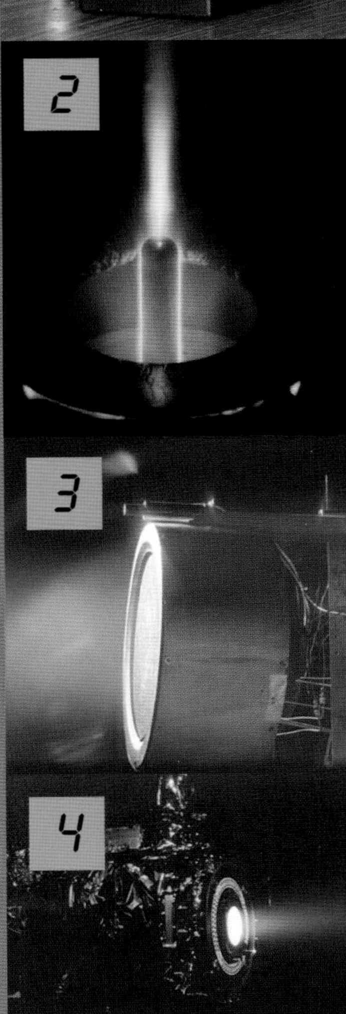

1 CHEMICAL ENERGY IS NOT SUFFICIENT TO PROPEL INTER-PLANETARY SHIPS ...

2 ... WHICH MUST USE VERY HIGH-PERFORMANCE ELECTRIC THRUSTERS ...

3 ... EJECTING CHARGED PARTICLES AT VERY HIGH VELOCITIES (ION MOTORS) ...

4 ... OR NEUTRAL CONDUCTING GAS (PLASMA MOTORS).

5 TO SUPPLY THE NECESSARY ENERGY, NUCLEAR ENERGY SOURCES ARE NEEDED ...

6 ... DISTANT DESCENDANTS OF THE NERVA ROCKET MOTORS FROM THE 1960S.

7 THE MOTOR CHOSEN FOR FLYING PEOPLE TO MARS IS THE REVOLUTIONARY VASIMR PLASMA THRUSTER.

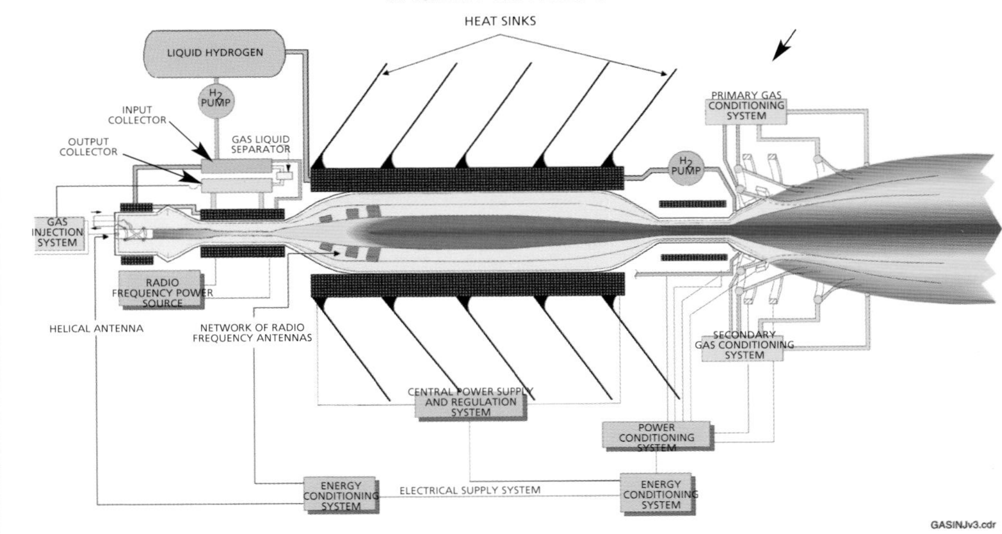

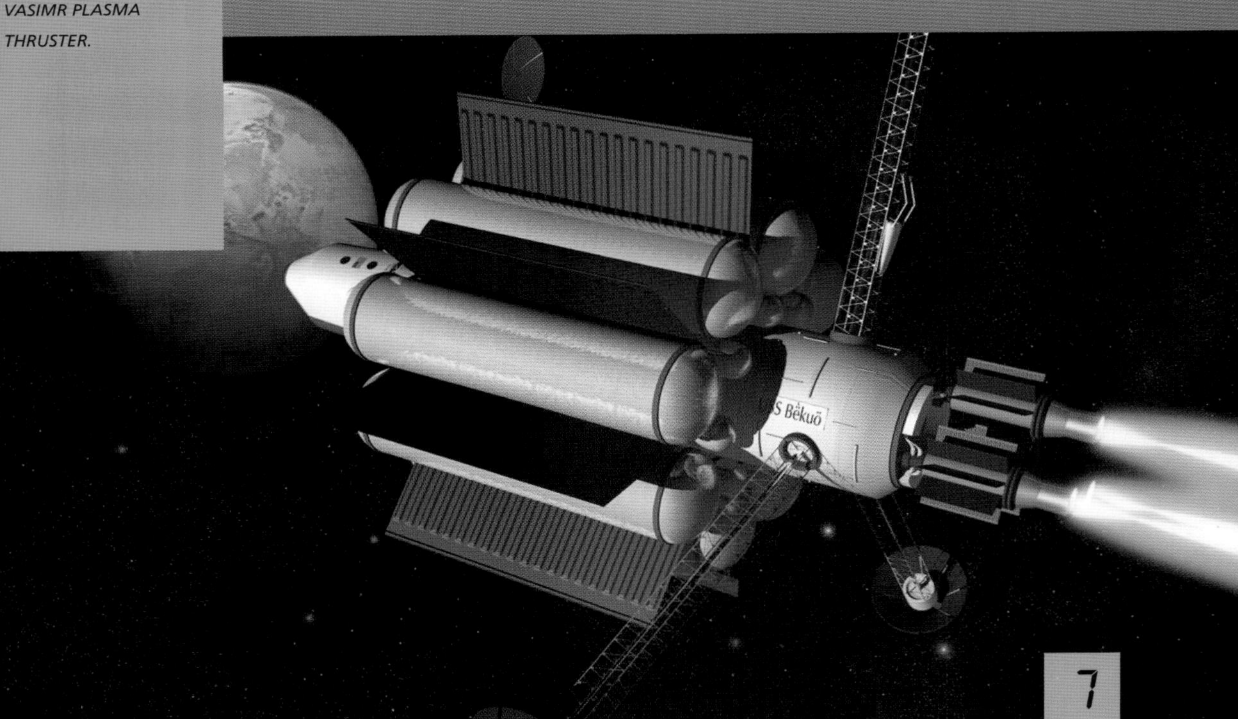

Nuclear Power in Space

The atom in space? This is not a new idea: During the Cold War, the United States and the Soviet Union placed nuclear reactors in orbit. Both compact and powerful, they provided a very convenient source of energy in space. Solar panels, on the other hand, are very large and at best supply 15,000 watts per square foot (1,400 watts per sq. m) of surface area; further, they do not operate in the shadow of the Earth or another body.

Nuclear power was not an option in the 20th century because of safety concerns. The first atomic reactors in space risked returning to Earth in an uncontrollable manner, which is exactly what happened to two Russian nuclear satellites, fortunately with minimal consequences. As well, no mission had required the high power provided by such energy sources.

Using solar energy was a problem with spacecraft like *Voyager*, *Galileo* and *Cassini-Huygens*, used for exploring distant planets. As they moved farther from the Sun, the power received from the Sun lessened, and the solar panels were less and less useful. At Mars, the power received from the Sun is already half as strong as it is near Earth; at the distance of Jupiter, it is 25 times weaker. The solution adopted for those spacecraft was a radio isotopic generator. This is not a nuclear reactor but a source powered by energy from the decay of radioactive isotopes. The available power is limited – a few hundred watts. A failure at launch is a risk, as the radioactive isotopes are dangerous and can be spread in the environment, whereas a nuclear reactor presents no risk during launch – it is not radioactive before starting off and the radioactivity is only produced far from the Earth.

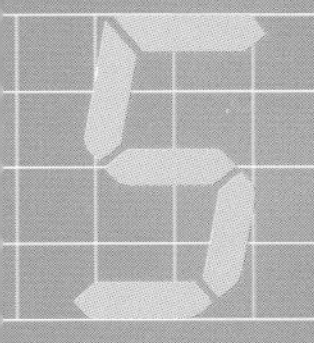

Forward to Mars

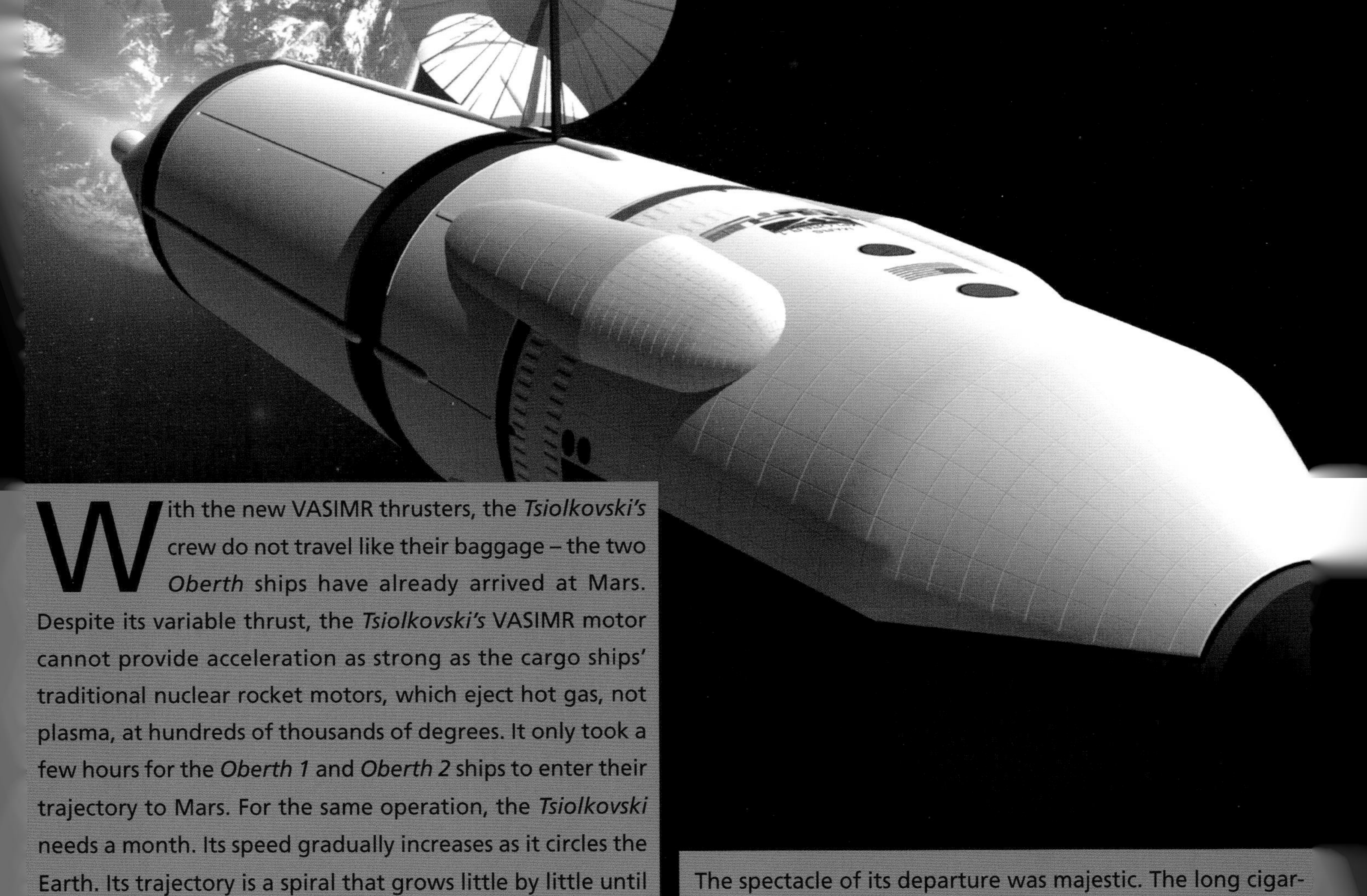

With the new VASIMR thrusters, the *Tsiolkovski's* crew do not travel like their baggage – the two *Oberth* ships have already arrived at Mars. Despite its variable thrust, the *Tsiolkovski's* VASIMR motor cannot provide acceleration as strong as the cargo ships' traditional nuclear rocket motors, which eject hot gas, not plasma, at hundreds of thousands of degrees. It only took a few hours for the *Oberth 1* and *Oberth 2* ships to enter their trajectory to Mars. For the same operation, the *Tsiolkovski* needs a month. Its speed gradually increases as it circles the Earth. Its trajectory is a spiral that grows little by little until it reaches the edge of the zone dominated by the Earth's gravity, about 625,000 miles (1 million km) away.

The presence of the crew on the *Tsiolkovski* during the ship's 30 days of spiraling would serve no useful purpose. Therefore, the seven astronauts watched the departure of their ship at the space flight control center in Houston, Texas. The *Tsiolkovski* was piloted by two astronauts – the ship specialists from the two alternate teams. The slow separation from the Earth was used to conduct a final series of tests intended to show that the *Tsiolkovski* was ready for its interplanetary journey.

The spectacle of its departure was majestic. The long cigar-shaped ship separated from the International Space Station under the power of small conventional rocket motors, accompanied by a flotilla of camera-bearing satellites that observed it from every angle. At 60 miles (100 km) from the station, the big moment finally arrived: the startup of the VASIMR thruster. A ball of extraordinary luminosity formed behind the ship, which started to accelerate majestically.

Godspeed ...

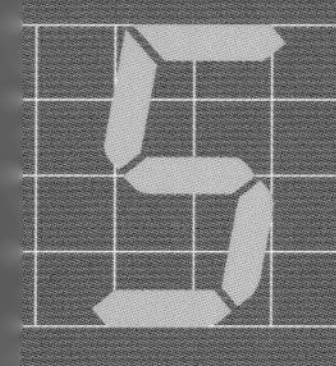

The Planned Itinerary ...

On March 16, 2033, the *Tsiolkovski* left the International Space Station (ISS), its base near the Earth. The crew for the Mars mission took off on April 4 from Cape Canaveral onboard the super shuttle *Sanger*, named in honor of the great German pioneer who first imagined a space plane. Its destination was the ISS, where a transfer ship was ready to take them to the *Tsiolkovski*, which was already outside the orbit of the Moon. The transfer ship is just one of the shuttles that completes regular round trips between the ISS and the lunar base. This time, a shuttle called *Kozlov* will pass the Moon and dock with *Tsiolkovski* 310,000 miles (500,000 km) from the Earth.

For the crew, the big departure toward Mars takes place April 5, 2033. The lunar shuttle *Kozlov* rapidly accelerates under the power of its chemical rocket motors. Those motors do not have the needed performance for a trip to Mars but are perfectly suited for journeys between the Earth and Moon. It takes four days to meet the Tsiolkovski, whose velocity continues to increase. It's time to catch up with it: In a few days, pushing forwards toward Mars, it will be beyond the reach of conventional spacecraft.

The time for the final goodbyes has come. The Mars crew takes over the Tsiolkovski, and the two pilots who directed the ship during its first three weeks board the lunar shuttle and follow the route back to Earth.

UNITED NATIONS

1

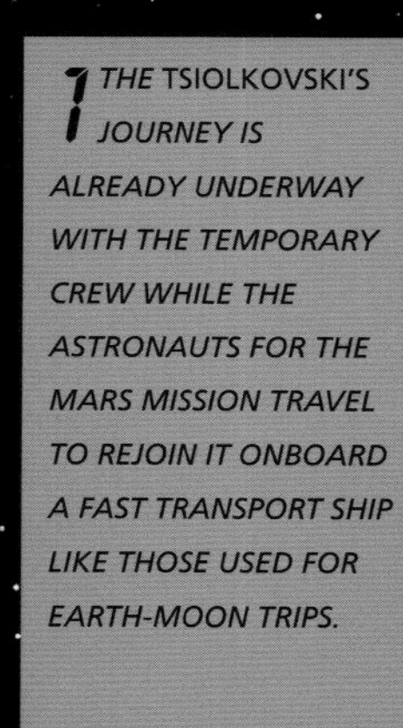

1 THE TSIOLKOVSKI'S JOURNEY IS ALREADY UNDERWAY WITH THE TEMPORARY CREW WHILE THE ASTRONAUTS FOR THE MARS MISSION TRAVEL TO REJOIN IT ONBOARD A FAST TRANSPORT SHIP LIKE THOSE USED FOR EARTH-MOON TRIPS.

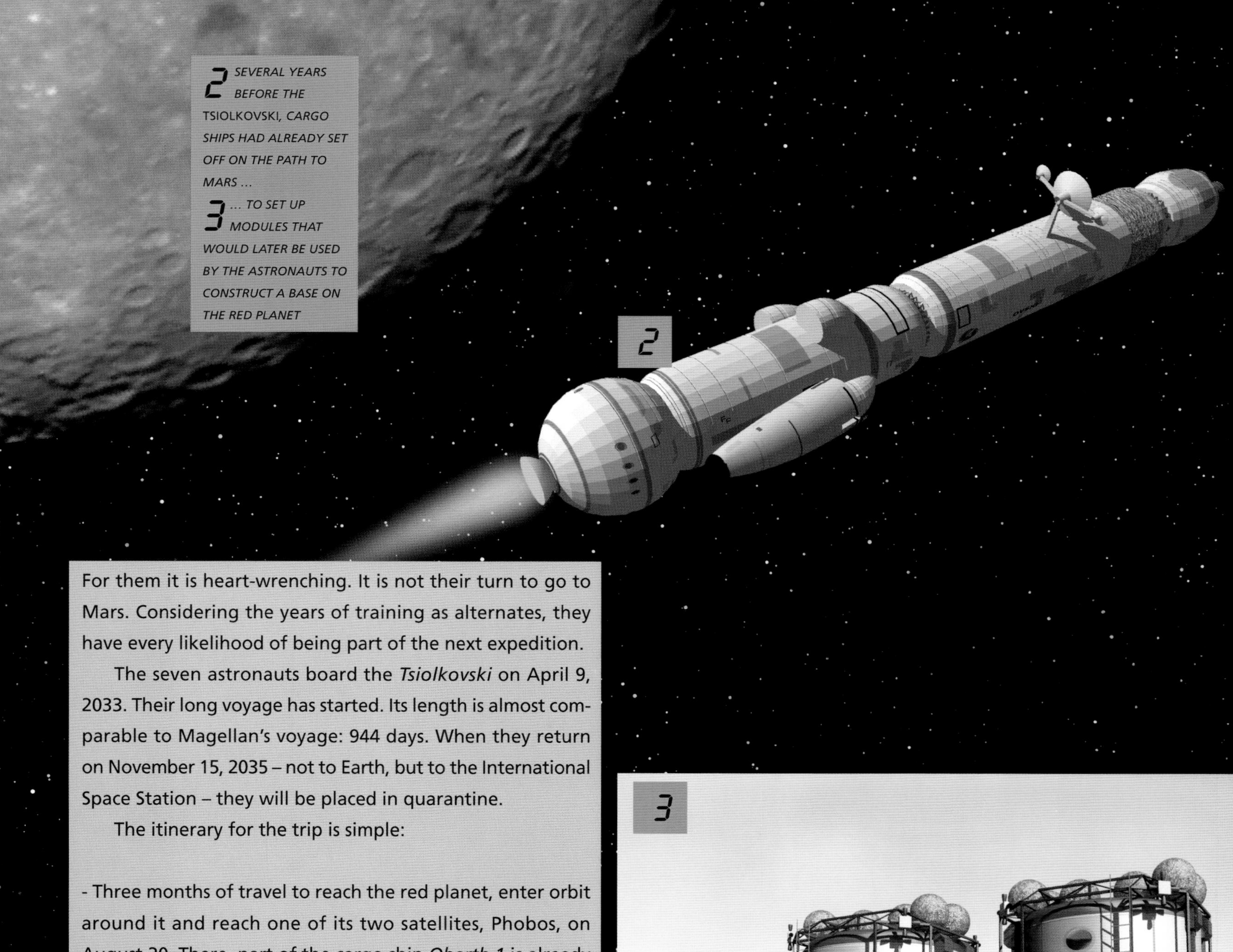

For them it is heart-wrenching. It is not their turn to go to Mars. Considering the years of training as alternates, they have every likelihood of being part of the next expedition.

The seven astronauts board the *Tsiolkovski* on April 9, 2033. Their long voyage has started. Its length is almost comparable to Magellan's voyage: 944 days. When they return on November 15, 2035 – not to Earth, but to the International Space Station – they will be placed in quarantine.

The itinerary for the trip is simple:

- Three months of travel to reach the red planet, enter orbit around it and reach one of its two satellites, Phobos, on August 20. There, part of the cargo ship *Oberth 1* is already moored;
- One and a half months of work on Phobos before descending to Mars and the touchdown, planned for October 4, 2033, on the Candor Chasma site where *Oberth 2* is already located;
- A total of 668 days on the red planet, until August 4, 2035, when they return to the base on Phobos and transfer to the part of the *Tsiolkovski* that is returning;
- A little over three months to return to the area near Earth, transfer to a lunar shuttle and, finally, return to the International Space Station on November 15, 2035;

A few weeks in quarantine then, finally, if everything goes well, a return to Earth in time for the Christmas holidays 2035.

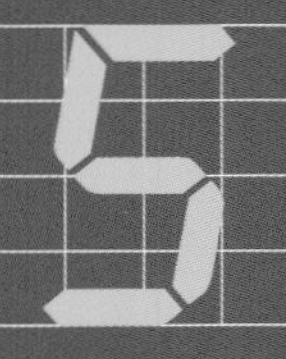

Into the Solar Wind

Accelerating on the way to Mars, the *Tsiolkovski* and its crew have left the Earth's protective cocoon, in particular, its magnetic field. Our planet is surrounded by a vast zone extending beyond the Moon, called the magnetosphere, in which the terrestrial magnetic field governs the movements of all the charged particles that traverse space. The magnetosphere plays a fundamental role in the protection of life on Earth. In its absence, the very high-energy particles ejected from the Sun would reach the Earth and subject plants and animals to a dangerous flux of radiation. It is a very effective screen; under its shelter, life was able to develop and prosper on the emerging lands. Of course, the screen is not completely sealed. Some solar particles do penetrate the magnetosphere, especially during periods of high solar activity that often follow an 11-year cycle. These particles, however, are guided toward the polar regions, where their arrival high in the atmosphere creates spectacular phenomena: the auroras.

The year 2033 is not a very favorable one for solar activity; it's far from the period of solar minimum, and the solar "storms" are numerous. During storm episodes, a very active zone appears on the Sun and is the source of a flux of particles, especially protons, that are very dangerous. On Earth, solar storms provoke serious disturbances in the propagation of short waves and even the distribution of electricity. In space, far from the shelter of the magnetosphere, the danger is fatal. Without protection, in just a few hours the flux of particles from the solar storm would subject the Tsiolkovski's astronauts to a lethal dose of radiation.

To avoid such a catastrophe, it is necessary to monitor the Sun. To detect active zones as early as possible, a network of spacecraft have been placed around the Sun, near the orbit of Mercury, in order to establish a kind of "meteorological map" and reveal the storms as they are forming. But the lead time can be short – only dozens of minutes. Therefore, it has to be possible to act quickly and move the astronauts to shelter. That is why the *Tsiolkovski* is equipped with a safe room, with thick metal walls and, more important, an inflatable shell filled with water, which offers the best protection against solar radiation. In case of an alert, the seven astronauts take their places in this safe room until the storm, which can last several days, passes. It isn't very comfortable, but it is indispensable.

But even with a safe room for solar storms, the Martian mission is not entirely free from radiation hazards. In normal times, solar particles are dangerous; in addition, there are other particles that are more energetic and much more penetrating: cosmic rays. This danger has been the principal reason for the work performed to shorten the trip between the Earth and Mars. On Phobos, as on Mars, the astronauts can spend a significant part of their time in a buried base. But in space, they are subjected to solar and cosmic radiation. The dose of dangerous radiation accumulated in a few months can exceed that authorized for the entire life of a worker in the nuclear industry. Therefore, there is a need to go fast – very fast. The three months of travel for the *Tsiolkovski* is, therefore, an immense improvement compared to six or nine months on more economical paths.

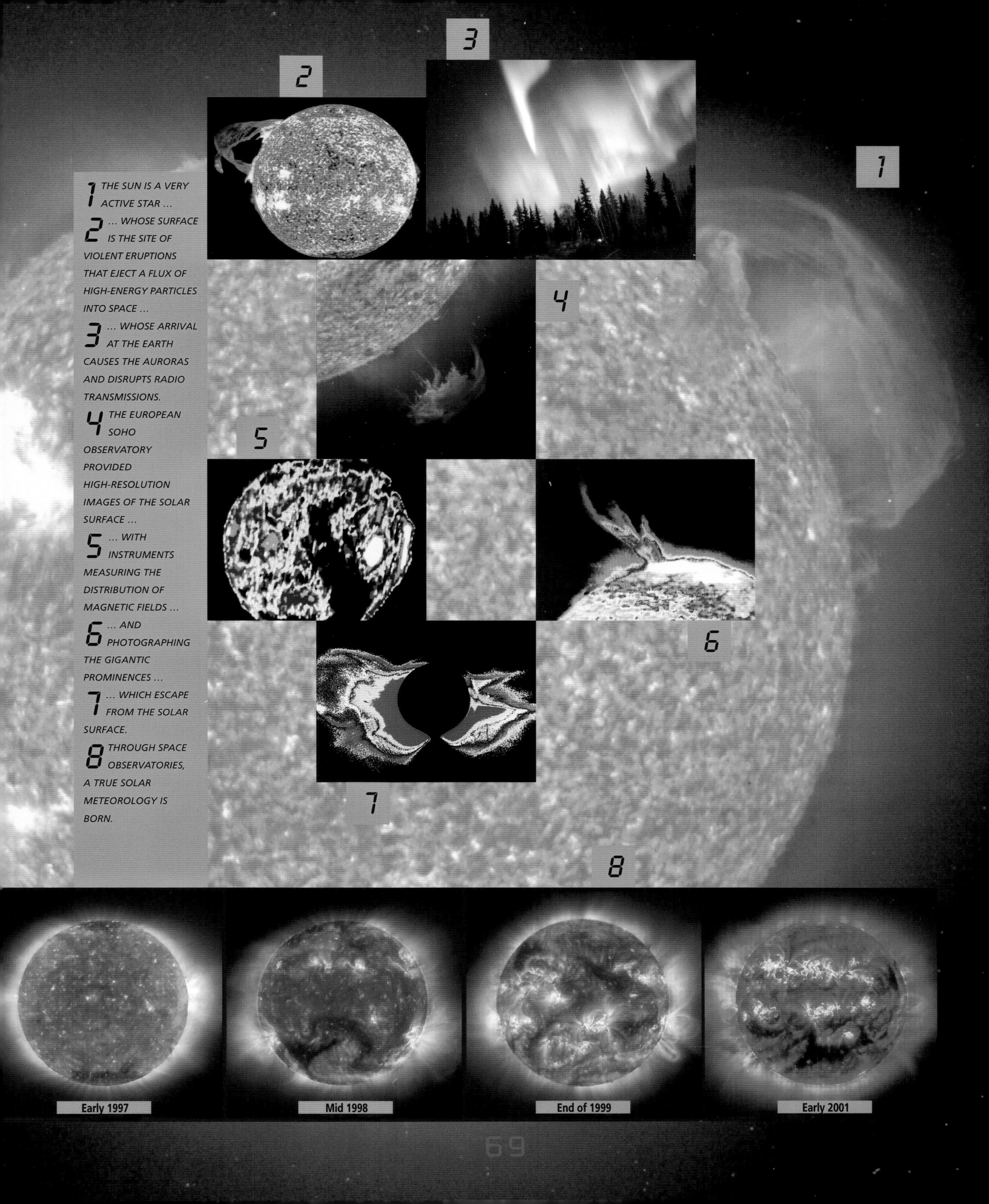

1 THE SUN IS A VERY ACTIVE STAR …

2 … WHOSE SURFACE IS THE SITE OF VIOLENT ERUPTIONS THAT EJECT A FLUX OF HIGH-ENERGY PARTICLES INTO SPACE …

3 … WHOSE ARRIVAL AT THE EARTH CAUSES THE AURORAS AND DISRUPTS RADIO TRANSMISSIONS.

4 THE EUROPEAN SOHO OBSERVATORY PROVIDED HIGH-RESOLUTION IMAGES OF THE SOLAR SURFACE …

5 … WITH INSTRUMENTS MEASURING THE DISTRIBUTION OF MAGNETIC FIELDS …

6 … AND PHOTOGRAPHING THE GIGANTIC PROMINENCES …

7 … WHICH ESCAPE FROM THE SOLAR SURFACE.

8 THROUGH SPACE OBSERVATORIES, A TRUE SOLAR METEOROLOGY IS BORN.

5 Meeting with an Asteroid

On July 14, 2033, an exceptional event disrupts the daily routine that had set in onboard the *Tsiolkovski*: a meeting with an asteroid. An asteroid is one of those bodies, generally rocky, orbiting the Sun like the planets, but whose dimensions are much more modest (several dozen yards to several hundred miles, for those that are worthy of being called a planetoid). Most asteroids are between the orbits of Mars and Jupiter in a region that the *Tsiolkovski* will not cross. But some of them move closer to the Sun than Mars and can even cross the orbit of the Earth, like asteroids in the Apollo family.

Some follow orbits that sometimes bring them very close to the Earth – the so-called "Near Earth Asteroids," or NEAs. Since the 1990s, they have especially interested scientists, who think that some of these asteroids were the cause of major catastrophes on our planet in the past, like the event that led to the disappearance of the dinosaurs 65 million years ago. Of course, NEAs are not the only possible suspects: comets, from the far reaches of the solar system, could also have crashed into the Earth. But if an NEA a few miles in diameter struck our planet, like in the film *Armageddon*, our civilization, even our species, and a large part of the life on Earth could disappear.

Detecting NEAs, predicting their approaches near the Earth, and thus the danger of cosmic collisions, has become a priority. For now, we are very fortunate that no NEA threatening to strike the Earth has been observed. But, all the same, the one that flies past the *Tsiolkovski* will pass only 60,000 miles (100,000 km) from the Earth in 2068. Called Asterix and discovered by a French observatory in 2020, it has a diameter of 3 miles (5 km). The path of the Tsiolkovski was chosen to fly by it only a few hundred miles away. The meeting is brief because the relative velocity of the asteroid and the ship is over 6 miles per second (10 km/s), but the sight of this dark, potato-shaped projectile, pockmarked with craters, is impressive.

But what if it threatens to strike the Earth? What could be done to turn away the cataclysmic threat? With the progress of space technology, it will perhaps be possible one day to modify the trajectory of a threatening celestial body. But we are not there yet. While we wait, there is only one means to guarantee the survival of the human species and civilization faced with such a risk: Set up an independent base on Mars or even make the red planet a second Earth. The voyage of the *Tsiolkovski* is a first step in that direction.

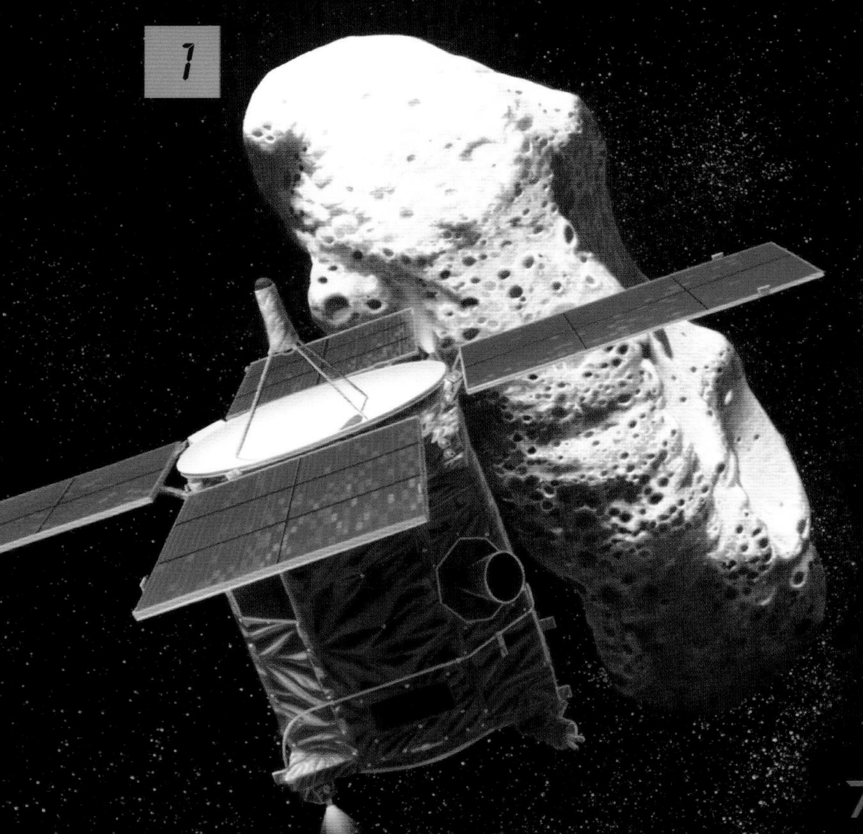

1 THE AMERICAN NEAR SPACECRAFT WAS THE FIRST TO GO INTO ORBIT AROUND AN ASTEROID, EROS, ON FEBRUARY 28, 2001.

2 AN ARTIST'S CONCEPTION OF THE ASTEROID TOUTATIS IN THE FOREGROUND, PASSING NEAR THE EARTH. TOUTATIS IS ONE OF THE NEAR EARTH ASTEROIDS WHICH REGULARLY PASS CLOSE BY OUR PLANET.

3 ON THE WAY TO JUPITER, THE AMERICAN GALILEO SPACECRAFT WAS THE FIRST, ON OCTOBER 29, 1991, TO FLY BY ONE OF THE ASTEROIDS THAT MOVE BETWEEN MARS AND JUPITER: GASPRA ...

4 ... BEFORE REPEATING THIS EXPLOIT AUGUST 28, 1993 WITH IDA, WHICH HAS A SURFACE COVERED WITH CRATERS AND A VERY ELONGATED AND IRREGULAR SHAPE: 14 MILES BY 36 MILES (23 KM BY 58 KM).

5 SOME ASTEROIDS HAVE UNDOUBTEDLY CAUSED ENORMOUS CATASTROPHES ON THE EARTH IN THE PAST.

Life on the Way to Mars

Apart from rare events, like a solar flare-up or a meeting with an asteroid, life on the *Tsiolkovski* is not very different from that familiar to all the members of the crew from their frequent stays on the International Space Station (ISS). There is an extraordinary freedom that the human body feels in the absence of weight. The VASIMR motor continuously exerts a small thrust, felt as light, artificial gravity. But this is only one one-thousandth of the Earth's gravity, and its only effect is to slowly move the astronauts who want to rest quietly suspended in the middle of the cabin.

Until the 2020s, life in microgravity was not without its inconvenience. It was accompanied by a regular loss of muscle mass and bone density, since the musculoskeletal system did not have to resist the body's weight. But with advances in genetic engineering, biologists have developed medicines that improve the absorption of calcium in the bones and maintain muscle mass, even in the absence of effort. The first beneficiaries of this progress were the elderly and the bedridden. But astronauts also benefit from these advances. Gone are the long hours of daily exercise on stationary bicycles or treadmills, on which the weightless athletes were held in place by elastic straps. The *Tsiolkovski*, like the ISS, has an exercise room, but it is only used for recreation and fitness.

There is still another consequence of weightlessness: cardiovascular deconditioning. The heart, arteries and veins no longer have to fight against the accumulation of blood in the lower part of the body, as happens on the Earth. Consequently, there is a reduction and redistribution of the volume of blood in circulation, which only takes a few days before the cardiovascular system adapts perfectly to microgravity. This deconditioning only poses a problem on the return of weight, first during braking maneuvers, and then on a celestial body – the Earth, Moon or Mars.

In a standing position, the blood moves to the lower part of the body, and the supply to the brain is no longer sufficient; the astronauts risk fainting. To avoid these dangerous symptoms on the return from a long mission on the ISS or a journey to Mars several months long, the astronauts must train for a return to weight. The exercise room includes a large centrifuge – a ring 26 feet (8 m) in diameter turning at a rate of one turn every 10 seconds and creating an artificial gravity equivalent to that on Mars (40 percent of the Earth's gravity). This ring is a running path on which it should be sufficient, in principle, to exercise for a few days before touchdown on Mars or the return to Earth. That is not a problem because running in artificial gravity is by far the crew's preferred exercise.

The trip from the Earth to Mars or, more specifically, from the edge of the Earth's zone of attraction to Phobos, is hardly any longer than the standard mission onboard the ISS – a little more than three months. With a crew of seven people, that represents a consumption of air, water and food of about 20 tons one-way and 40 tons for the round trip. That is far from being negligible compared to the 300 tons of the *Tsiolkovski* on departure from the Earth. To reduce this mass, the solution, implemented by the Russians in their Salyut orbital stations in the 1970s, is simple: Recover the water given off by the astronauts by various means (urine, sweat and stool), purify it and redirect it into the circuit. The food that is transported is dehydrated; this measure makes it possible to reduce the mass necessary for supplying the crew by two-thirds, which is entirely acceptable. On Mars, where the astronauts will stay for two years, another, much more elaborate solution will be used: the creation of a true closed ecological system, like that which supports life on "spaceship Earth."

1 THE IMMENSE GREENHOUSE BIOSPHERE 2 WAS CONSTRUCTED IN THE ARIZONA DESERT TO EXPERIMENT WITH A CLOSED ECOLOGICAL SYSTEM; THAT IS, LIFE IN A CLOSED CIRCUIT, WITH PEOPLE, PLANTS AND ANIMALS, AND COMPLETE RECYCLING OF ALL WASTE.

2 ON BOARD ORBITAL STATIONS, THE ASTRONAUTS ALREADY LARGELY RECYCLE WATER, BUT IT'S ONLY THE BEGINNING. ON THE TSIOLKOVSKI, MOST OF THE RESOURCES ARE RECYCLED.

3 THERE IS ONLY ONE WAY TO STAY IN SHAPE: EXERCISE (HERE ON A TREADMILL).

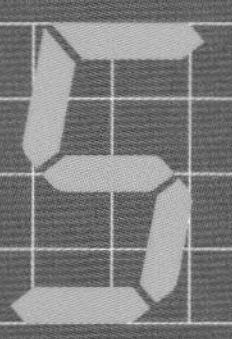

Meeting with Phobos

The days pass, still on the schedule of the principal control center in Houston, Texas, but little by little the astronauts' time is no longer that on Earth. Every 625,000 miles (1 million km) of separation adds six seconds to the round trip time for signals between the blue planet and the *Tsiolkovski*. At the beginning of the voyage, multiple televised connections gave the astronauts the impression of still being on Earth. But by the halfway point, on August 1, 2033, the Earth is already 22 million miles (36 million km) away; and almost four minutes pass between a question and its answer. Direct conversations are no longer possible; messages on computer screens replace discussions. The crew is alone before its mission, in the light of the shrinking Sun and already far from the Earth, which is the brightest "star" in the firmament, bluish against the black sky, and far from another "star" that is a little less brilliant but red in color, Mars.

The solitude is the most novel experience of the mission and distinguishes it from all previous space flights. Even on the Moon, you're only one light-second from the Earth. Several minutes is entirely different. In case of a problem, the crew can only count on its own abilities; often the control center only hears about difficulties after they have been resolved. The remoteness increases the crew's solidarity. But the approach of Mars restores motivation to all of the astronauts. On September 15, the *Tsiolkovski* enters the red planet's zone of attraction, and its VASIMR motor changes speed; now it's a matter of gently slowing the ship's movement to bring it progressively into orbit. This maneuver is exactly the opposite of that performed to leave the Earth. The *Tsiolkovski* slowly traces an increasingly tighter spiral around Mars. Its objective is not a space station but a natural satellite of the red planet, Phobos, which turns 5,800 miles (9,400 km) above Mars' equator.

Phobos is an ideal stopping point. With modest dimensions – 11 by 14 by 18 miles (18 by 23 by 28 km) – its gravity is very weak, and approaching it does not require any use of energy. Phobos was probably an asteroid captured in the distant past by the gravitational attraction of Mars. Reflecting little light and having an irregular shape similar to that of a potato, it is a carbonaceous chondrite type asteroid, containing about 20 percent water in the soil along with molecules containing carbon. That makes it an extraordinary source of material for the Tsiolkovski's mission. In some ways, it's a service station on the road to Mars.

The heavy ship makes its historic meeting with the Martian satellite on September 25, 2033. Harpoons shoot out and stick fast in the ground like anchors from naval ships. The cables grow taut and the feet deploy. Soon the *Tsiolkovski* is firmly moored on the surface of Phobos a few hundred yards from another, even larger ship, Oberth 1, that arrived two years earlier.

1

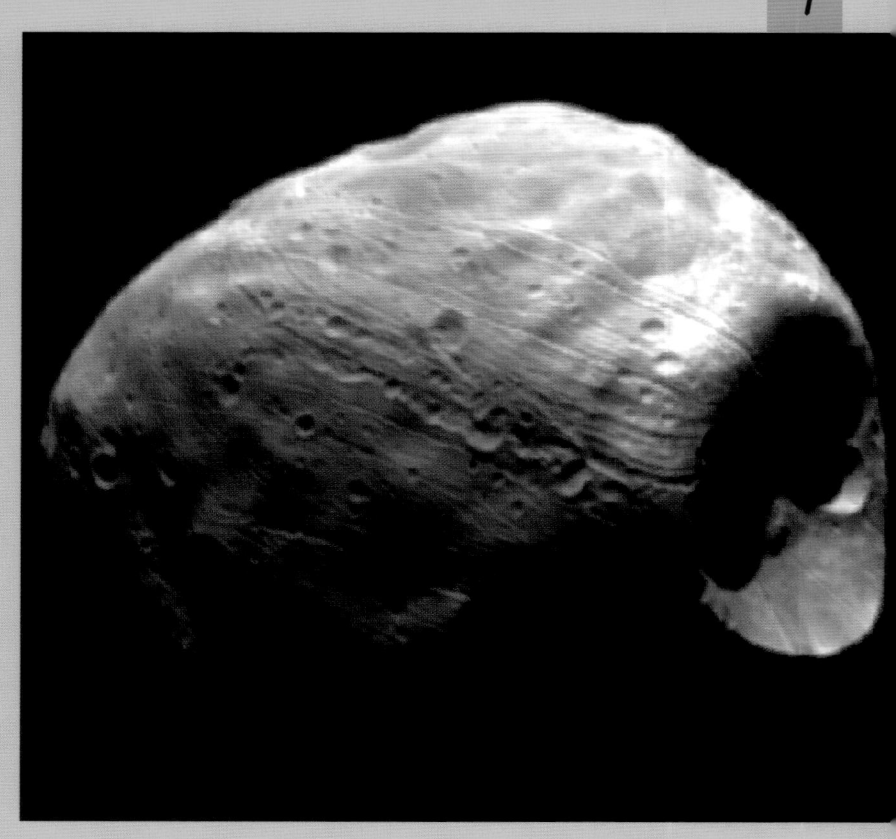

1 PHOBOS, MARS' LARGER SATELLITE, WAS PHOTOGRAPHED IN DETAIL BY THE VIKING SPACECRAFT.

2 THE ASTRONAUTS STOP THERE BEFORE DESCENDING TO MARS, TO INSTALL AN INTERMEDIATE BASE AND MINE ITS RESOURCES. THE VIEW OF THE RED PLANET IS SPECTACULAR.

5 Mars Seen from Above

If the Earth had a satellite like Phobos, both near and small, instead of the Moon that is distant and nearly as large as Mercury, then interplanetary voyages would have been much more accessible and there would have been no need to construct orbital stations. That satellite would have played the role perfectly, just like Phobos is a fantastic base camp right next to Mars. The red planet's low mass compared to the Earth's would have also made departures into space easier. In other words, the conquest of space would have been much simpler for Martians than for us.

The view of Mars from Phobos is imposing. The ocher disk of the planet occupies a large quarter of the sky – 80 times larger than that of the Earth seen from the Moon. The equatorial zones of the red planet parade rapidly by the astronauts' eyes; Phobos completes three revolutions of the planet in one Martian day. The length of the day on Mars is called a "sol," and is 37 minutes longer than the Earth's. As soon as the *Tsiolkovski* moored at Phobos, the crew changed clocks and adapted to the Martian day. Each day, though, time onboard the ship slips compared to that in Houston. The different rhythm of days and nights is now added to the effects of the distance. The astronauts actually live in a different world, where the least exchange takes eight minutes now and will reach about 40 minutes when Mars and the Earth are on opposite sides of the Sun in August 2034.

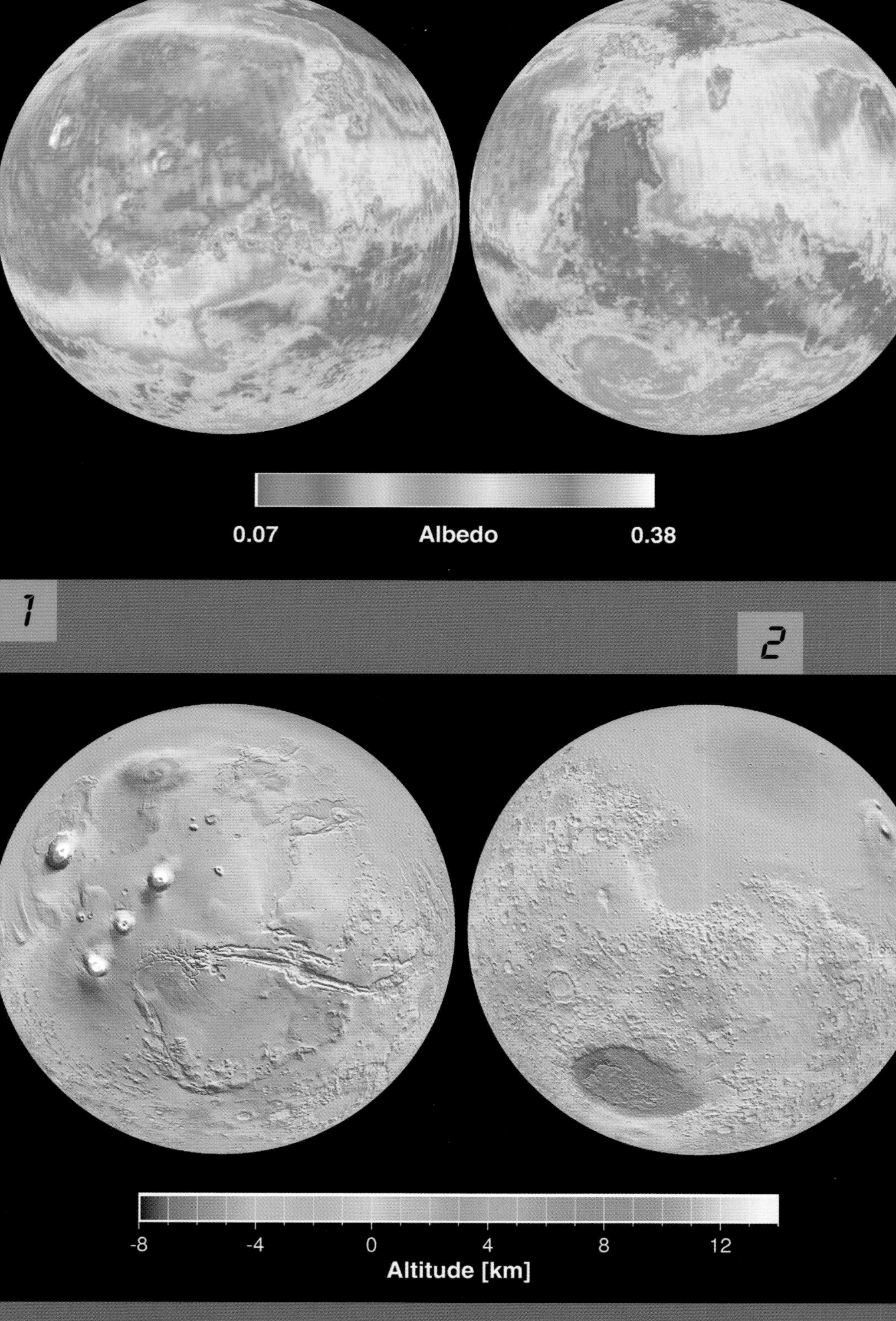

At least once per day, or rather per sol, Phobos passes nearly directly over the most extraordinary formation on the red planet, Valles Marineris. The immense feature, 2,500 miles (4,000 km) long, lies nearly parallel to the equator, just a little south of it. The flight over Valles Marineris takes over an hour and a half, during which the astronauts have difficulty tearing themselves away from the spectacle and accomplishing the fundamental task that they must perform: preparing to explore Mars and the return trip. Working on Phobos is both easy and difficult. It is easy because the gravity is a thousand times weaker than on Earth, so the astronauts can easily transport loads of several tons. It is difficult because a small kick is sufficient to send the astronauts into space; spacesuits, equipped with individual rocket thrusters, must remain firmly attached by cables to Phobos, the two ships moored to the asteroid, or to the base set up on the edge of the vast Shockley crater.

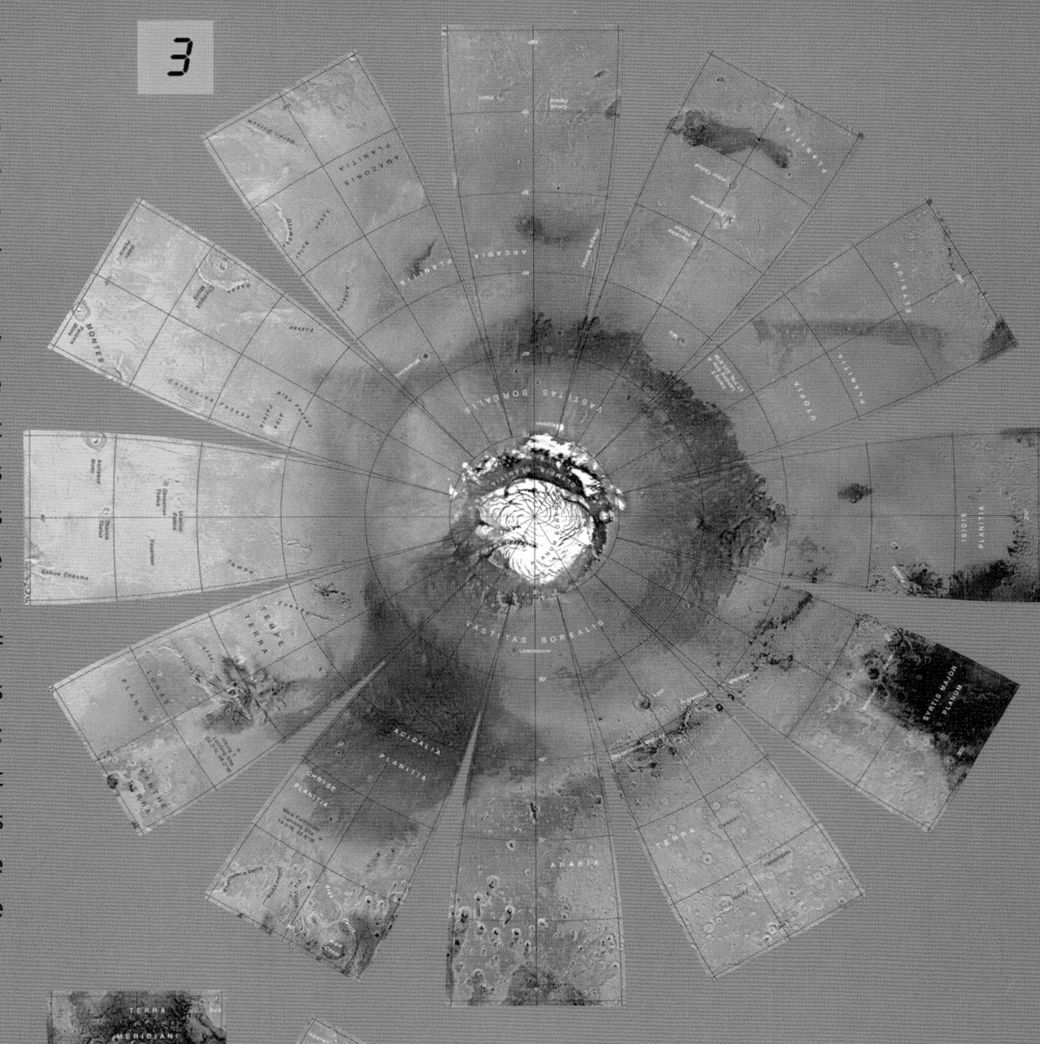

4

1 WITH MEASUREMENTS FROM MARS GLOBAL SURVEYOR, *VERY PRECISE MAPS OF MARS HAVE BEEN MADE BOTH FOR ITS SURFACE REFLECTIVITY (ITS "ALBEDO") ...*

2 ... AND FOR THE ALTITUDE OF ITS RELIEF.

3 THE NORTHERN HEMISPHERE OF THE RED PLANET HAS A MUCH LARGER POLAR CAP ...

4 ... THAN THE SOUTHERN HEMISPHERE.

5 In Orbit around Mars

The first people on Phobos initially turn to the work of moving in. They extract cargo from *Oberth 1* – equipment indispensable to pursuing the mission and, in particular, the elements of the factory that will extract materials from under Phobos' surface and fabricate needed propellants. The heart of this installation is the 1 MW nuclear reactor, which will furnish the necessary energy for the base installed on the Martian satellite and for the processing of carbonaceous and hydrated rocks. It is linked to a drilled well, which will seek material several dozens of feet deep to avoid disturbing the asteroid's surface, and a small chemical factory. The factory will extract water and carbon present in the soil and use these basic substances to produce oxygen, hydrogen and methane.

The hydrogen will be liquefied and stored on Phobos before filling the *Tsiolkovski*'s tanks for the return to Earth. The oxygen and methane will serve for the propulsion of the landing vehicles that the astronauts will take to descend to the red planet. That vehicle was brought to Phobos by the *Tsiolkovski*, a second unit is on board the *Oberth 1*. It is kept in reserve, ready to descend in case of need, to look for all or part of the crew at any point on Mars. A third landing vehicle, a cistern type, was also placed at the work site by *Oberth 1*. Its mission will be to descend to Mars toward the end of 2033 with about 10 tons of water produced on Phobos; this water will serve both directly for the life of the crew and as a source of oxygen. It would have been simpler to look for water in the subsoil of Candor Chasma, where the Martian base will be installed, but the robots that already explored that area showed that if water was present, it would be too deep to easily take advantage of it.

A final task must be performed before leaving Phobos: Several dozen small mobile robots will be sent to the sites on Mars that are candidates for a visit by the astronauts. These robots take advantage of the latest progress in miniaturization and computer science and have a mass of under 110 pounds (50 kg). But, supplied with energy from small nuclear electricity sources, they're capable of crossing several dozens of miles of uneven terrain, climbing steep slopes and even jumping crevasses. Through a network of communication satellites already in use around the planet, they will retransmit images of their journey to the Martian base in real time and can be remotely guided by the astronauts.

As the middle of November approaches, everything is ready. The landing vehicle's tanks are filled. The latest tests are complete. The base on Phobos is put to sleep. The astronauts will only make a brief stop there on their return. But it will from now on be a required stopover point for all conquerors of the red planet.

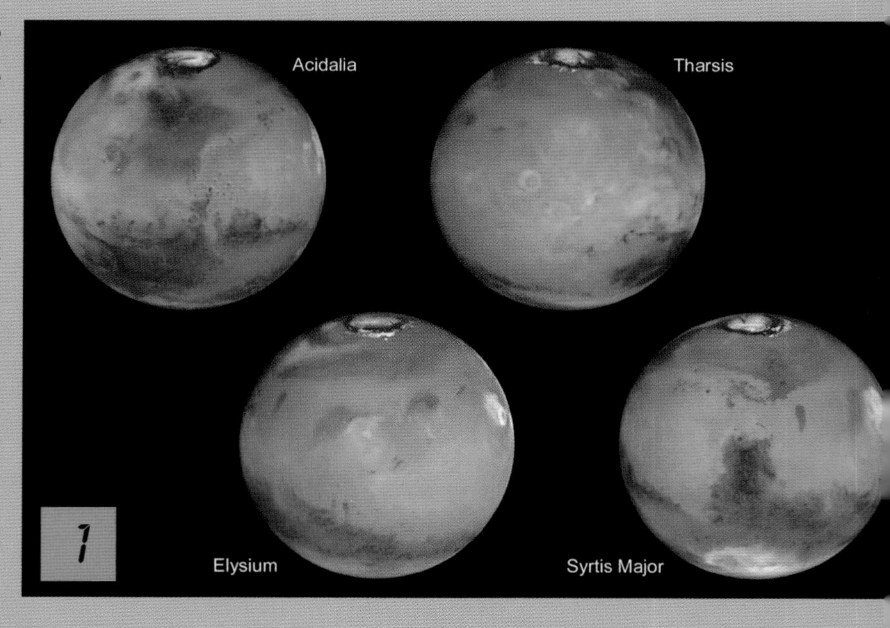

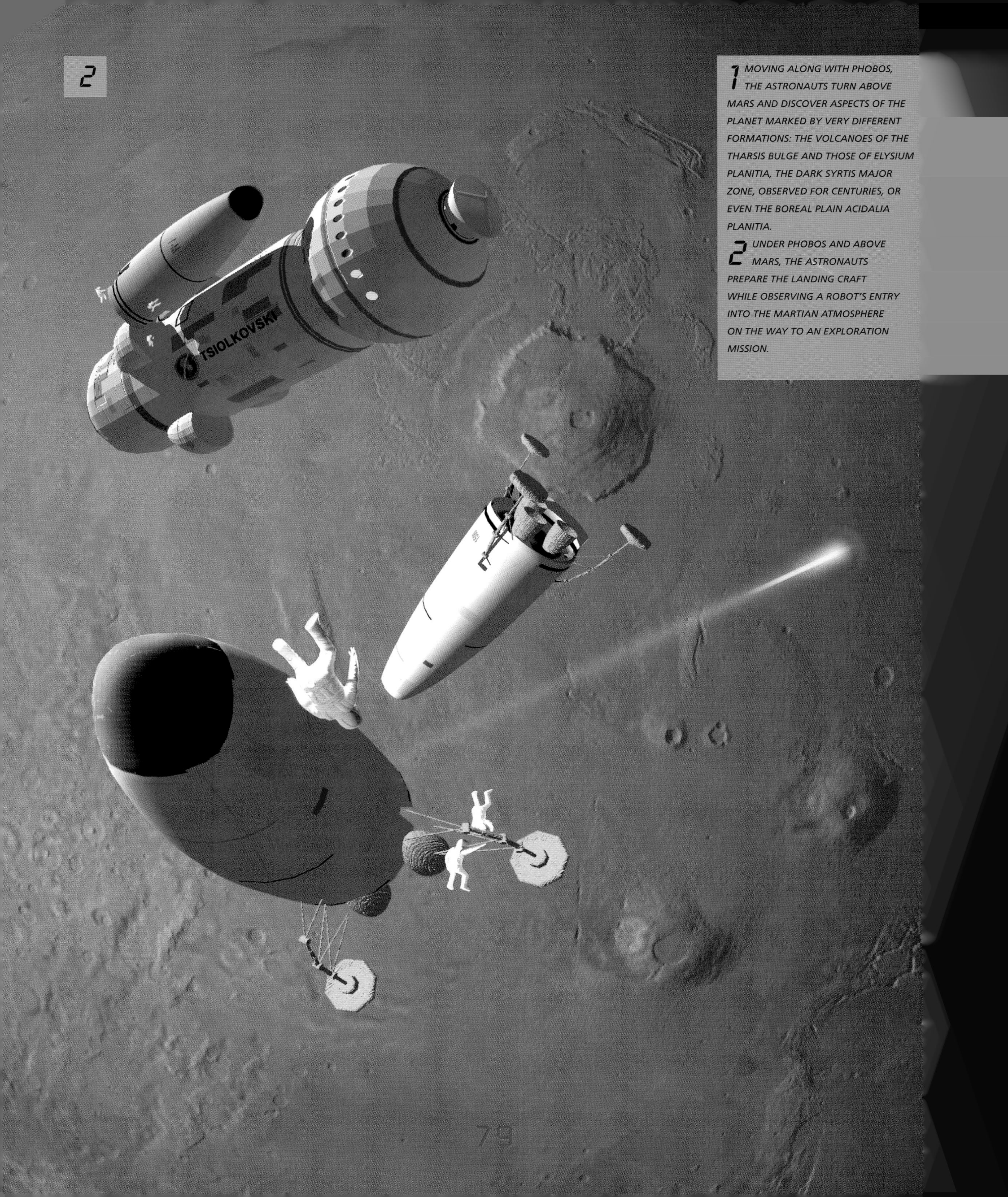

1 MOVING ALONG WITH PHOBOS, THE ASTRONAUTS TURN ABOVE MARS AND DISCOVER ASPECTS OF THE PLANET MARKED BY VERY DIFFERENT FORMATIONS: THE VOLCANOES OF THE THARSIS BULGE AND THOSE OF ELYSIUM PLANITIA, THE DARK SYRTIS MAJOR ZONE, OBSERVED FOR CENTURIES, OR EVEN THE BOREAL PLAIN ACIDALIA PLANITIA.

2 UNDER PHOBOS AND ABOVE MARS, THE ASTRONAUTS PREPARE THE LANDING CRAFT WHILE OBSERVING A ROBOT'S ENTRY INTO THE MARTIAN ATMOSPHERE ON THE WAY TO AN EXPLORATION MISSION.

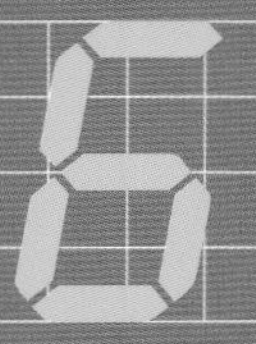

Touchdown

The great day, November 15, 2033, has come. The seven astronauts take their places in the landing vehicle that leaves Phobos, slowly moving away. Its engines ignite to greatly reduce its velocity – from a little over half a mile (1 km) per second – and place the vehicle on a trajectory that will have it move from 5,800 miles (9,400 km) above Mars to only a few hundred miles in just three hours. Then it will penetrate the planet's atmosphere, which will progressively slow the lander through the action of aerodynamic friction. The deceleration is weak – not more than twice the strength of gravity on Earth. For the crew, who continued training in the centrifuge on the *Tsiolkovski*, this is the big return of body weight.

The landing craft has the shape of a "gliding body"; this allows it to maneuver in Mars' atmosphere and direct itself towards its objective: Candor Chasma. The last part of the descent is spectacular – a long, gliding flight above Valles Marineris, including Candor Chasma, one of its many canyons. But the red planet's atmosphere is not sufficiently dense to allow this gliding to continue until touchdown. The vehicle must first deploy a gigantic parachute that will sufficiently slow it to limit the magnitude of the final braking maneuver, which is performed by the vehicle's rocket engines.

The precision of the landing is impressive: The vehicle sets down within 30 feet (10 m) of the intended spot. For several years now, Mars, like the Earth, has had a network of navigation satellites that use beacons to determine a location on the planet, in its atmosphere or in space nearby, with the precision of a few yards. And so, arrival locations that are free of danger can be chosen for automatic ships just like inhabited vehicles. This also makes it possible for explorers – robotic or human – to very accurately know where they are.

The surface of Mars is like a desert. But even so, the astronauts do not arrive in a completely barren landscape. Mars is already well equipped with satellites for communication and navigation, and the *Oberth 2* cargo ship that brought the equipment necessary to build the von Braun base.

IV
Living on MARS

Candor Chasma II

The site for the base was carefully chosen in a flat region, called Candor Chasma II, at the bottom of the immense Valles Marineris canyon. On Earth, this location would be described as being located in a tributary of the canyon. But on Mars? Was there ever a river in the middle of this canyon? The mission of the Mars 1 crew is to discover just that. Still, the site is magnificent with the 1.2-mile (2 km) high canyon walls on the horizon. The scientists hope they will act like an open book of Mars' history. In a few weeks, an expedition will leave for the canyon walls on foot, and some astronauts may try to climb them if the risk isn't too great.

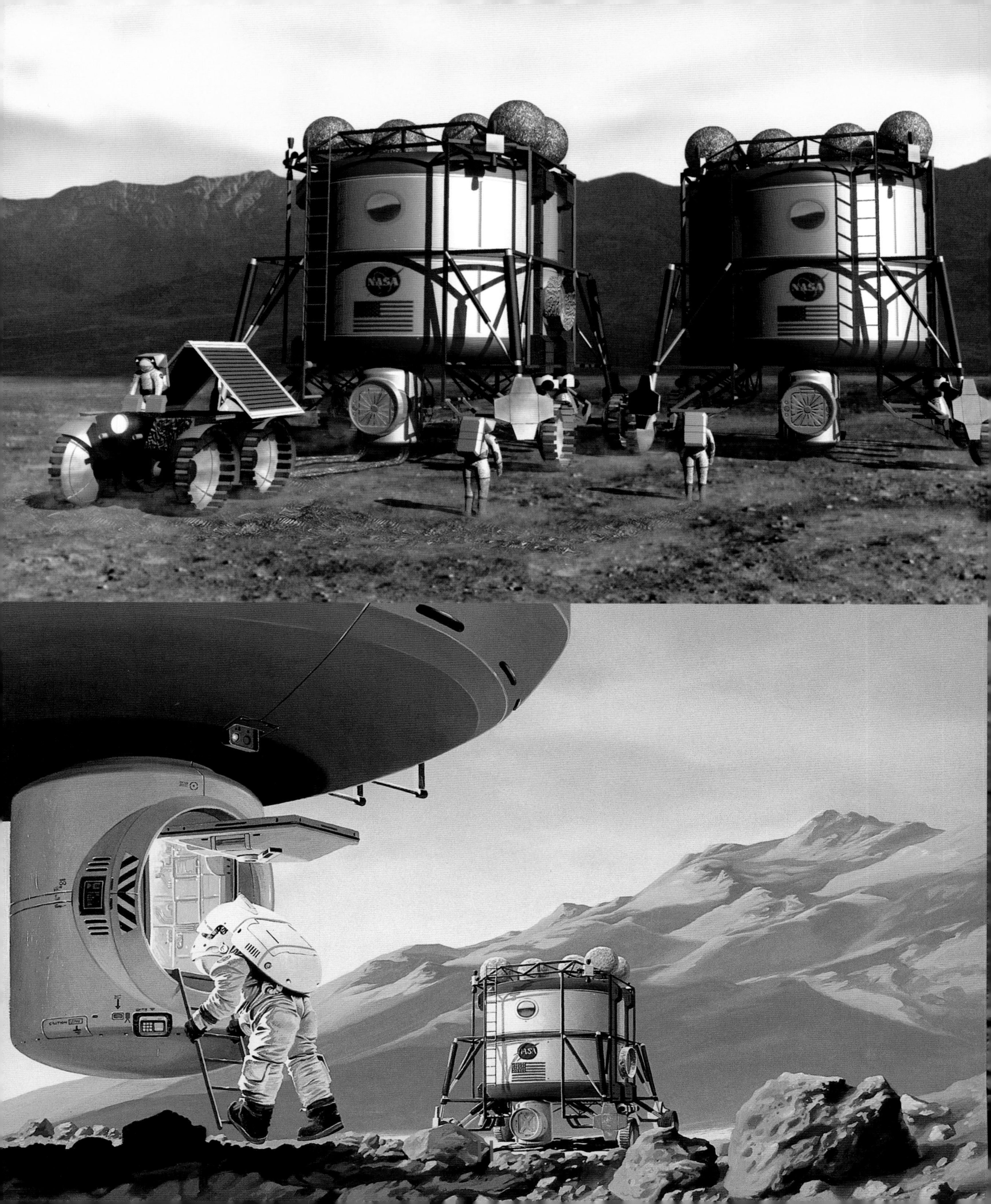
NASA
NASA

The Resources on Mars

The seven astronauts are going to spend nearly two years on the red planet. Can they get by, in part, on resources found locally? In the equatorial regions where the expedition is setting up, the possibilities are limited by the absence of an easily accessible source of water nearby. That is why the itinerary for the trip called for setting up an intermediary base on Phobos where water and carbon could be extracted. Some of the water obtained in that way will be transported to Mars, where it will be put to several uses: increasing the reserves of water for use by the crew during their daily existence and supplying oxygen by electrolysis. The astronauts will breathe some of the oxygen; the rest will be used as fuel for the Martian vehicles' motors and the thrusters on the rocket on which the astronauts will leave Mars (the complementary fuel, methane, will also be imported from Phobos, where it is manufactured from carbon and hydrogen).

The cistern ship left on Phobos will make several round trips between the von Braun base and the moon, bringing dozens of tons of water and methane each time. The solution is complex and the leaders of the Mars exploration program want to take advantage of the Mars 1 expedition to test other methods of taking advantage of the planet's resources. Although tenuous, the atmosphere is a promising source; it contains mostly carbon dioxide gas, whose molecules can be broken apart by chemical methods to supply oxygen and carbon. It also contains nitrogen, which could be used to compensate for the loss of this gas that makes up three-quarters of the atmosphere in the living quarters. (Their air has the same composition as on Earth.) And, finally, it also includes water vapor, although at very low concentrations, but it could nonetheless be extracted and supplement the supplies from Phobos. The astronauts are therefore setting up a small experimental factory to take advantage of the Martian atmosphere. The factory is supplied with energy by the three nuclear reactors on the *Oberth 2.*

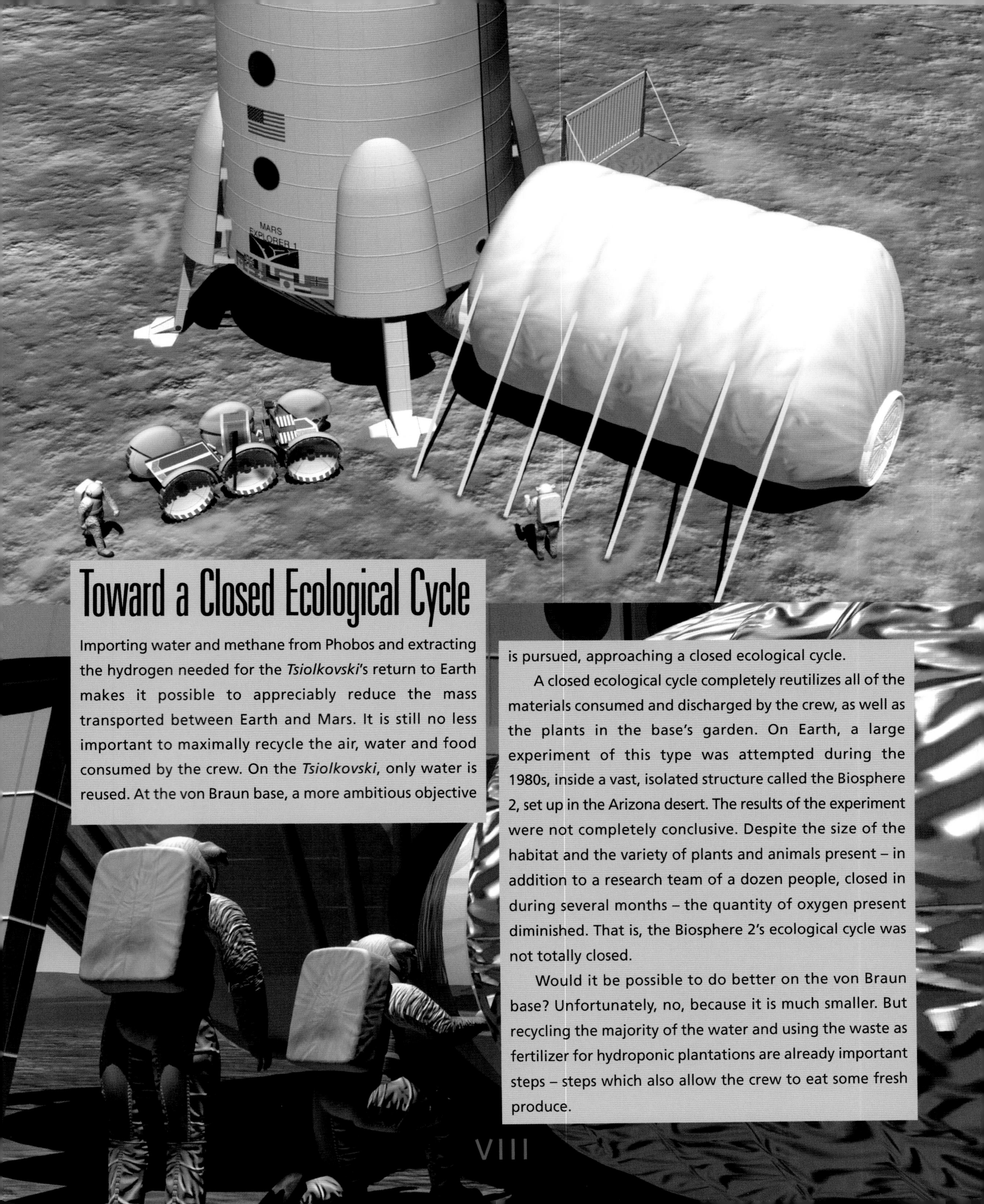

Toward a Closed Ecological Cycle

Importing water and methane from Phobos and extracting the hydrogen needed for the *Tsiolkovski*'s return to Earth makes it possible to appreciably reduce the mass transported between Earth and Mars. It is still no less important to maximally recycle the air, water and food consumed by the crew. On the *Tsiolkovski*, only water is reused. At the von Braun base, a more ambitious objective is pursued, approaching a closed ecological cycle.

A closed ecological cycle completely reutilizes all of the materials consumed and discharged by the crew, as well as the plants in the base's garden. On Earth, a large experiment of this type was attempted during the 1980s, inside a vast, isolated structure called the Biosphere 2, set up in the Arizona desert. The results of the experiment were not completely conclusive. Despite the size of the habitat and the variety of plants and animals present – in addition to a research team of a dozen people, closed in during several months – the quantity of oxygen present diminished. That is, the Biosphere 2's ecological cycle was not totally closed.

Would it be possible to do better on the von Braun base? Unfortunately, no, because it is much smaller. But recycling the majority of the water and using the waste as fertilizer for hydroponic plantations are already important steps – steps which also allow the crew to eat some fresh produce.

6 In Spacesuits on Mars

Even at the bottom of Candor Chasma II, which is located well below the planet's average altitude, the Martian air is extremely tenuous. Its pressure is 100 times less than the Earth's atmosphere at sea level. In this respect, isn't the red planet more like the Moon, where a total vacuum reigns, than the Earth? No, as the presence of an atmosphere, even a very thin one, changes everything. The sky is not black, with brilliant stars at midday, but a beautiful salmon color, with an occasional light cloud. On winter nights, a small layer of frost is deposited on the surface. The micrometeorites do not come all the way to the ground but burn up as shooting stars, like they do on Earth. All this does not give the impression of being in space, but really and truly in another world – humanity's second Earth.

Even so, it's not possible to go out without spacesuits. For the first steps taken on the red planet, the astronauts used traditional spacesuits, with a rigid body and helmet, and flexible parts, containing an interior atmosphere of pure oxygen under reduced pressure. The spacesuits, which are useful for walks in empty space on the Moon or Phobos, are very uncomfortable and inconvenient. They have to be put on two hours before going out so the body can get rid of nitrogen dissolved in the blood and adapt to breathing pure oxygen. They're also heavy and inflexible.

For the Mars 1 expedition, another type of spacesuit has been developed. Aside from the helmet, it's like a second skin, and it directly exerts the necessary pressure on the body without an intermediate layer of air. Americans call it a "skinsuit." In fact, these clinging spacesuits are made up of several "skins" with circulating oxygen and water, and insulation to maintain the body at the right temperature. Each suit is supplemented by a backpack, with reserves of air and water, and communication and navigation systems. Of course, it is still a spacesuit. But with a weight two and a half times less than on the surface of the Earth, the "skinsuited" astronauts feel a sensation of comfort and freedom.

What's the Weather Like on Mars?

Protected by their living units, or spacesuits, the astronauts are largely insensitive to Martian meteorology, which somewhat resembles terrestrial deserts, with enormous temperature variations between night and day. The first steps on Mars took place a little south of the equator during the southern summer, when the daytime temperatures are their highest – they can reach 85°F (30°C). But don't be fooled by this value, as it is exceptional. The nighttime temperature drops down to –110°F (–80°C). These rapid variations occur because the red planet's atmosphere is very tenuous and does not retain heat through the greenhouse effect like the Earth's atmosphere. It's also nearly free of clouds (during the day, when the heat causes the Martian air to rise, only small convective clouds form at high altitudes) and does not rain. If all the water contained in the atmosphere actually fell to the ground, it would form a layer less than a tenth of a millimeter deep.

When seen from Mars, the Sun's diameter appears twice as small as when it's seen from Earth. The light is less intense; but without water vapor and atmospheric clouds, the day seems fully as bright as on the blue planet. The sky is pinkish gray before sunrise and then takes on a salmon color due to the presence of red dust particles at high altitudes in the atmosphere. The wind generally does not blow more than 12 to 18 mph (20 to 30 km/h), and considering the low density of the air, the force that it exerts on the astronauts, vehicles and the base's buildings is entirely negligible. The Martian winds still have a very important role: They lift up minuscule particles of dust from the surface and carry them to high altitudes. Sometimes the effect is spectacular, as when they form tornadoes, and especially when they form large dust storms, which are the most remarkable atmospheric events on the red planet.

These storms generally start during the southern summer, when Mars is closest to the Sun and the temperature difference between the equator and the South Pole is less pronounced. A large depression can then form in the Hellas Planitia, the largest plain on Mars, and give rise to a tempest carrying enormous quantities of dust into the sky. This dust storm can extend over the entire planet, masking the surface behind a red veil, and hiding the Sun behind a reddish obscurity. This phenomenon is what prevented *Mariner 9* from observing the Martian surface for two months in 1971. The storms are also responsible for changes in the surface's appearance because of the transport of dust from one region to another. In the 19th century, these color variations had been incorrectly interpreted as signs of plant life.

1

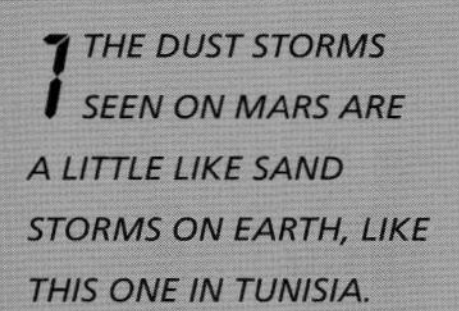

1 THE DUST STORMS SEEN ON MARS ARE A LITTLE LIKE SAND STORMS ON EARTH, LIKE THIS ONE IN TUNISIA.

2 THE SEASONS HERE ARE MORE MARKED THAN ON EARTH BECAUSE OF THE LARGE VARIATIONS IN THE DISTANCE BETWEEN THE SUN AND MARS DURING THE 687 DAYS OF THE MARTIAN YEAR.

3 A LARGE DUST STORM LEAVING THE MARTIAN NORTH POLE (ABOVE) RESEMBLES A SAND STORM STARTING AT THE SAHARA AND SPREADING WEST OF AFRICA (BELOW).

4 FROST ON UTOPIA PLANITIA PHOTOGRAPHED BY THE VIKING 2 LANDER ON MAY 18, 1979. THE THICKNESS OF THE ICE IS LESS THAN A FEW HUNDREDTHS OF A MILLIMETER.

5 HELLAS PLANITIA, BELOW, IS COVERED WITH FROST IN THIS MOSAIC OF IMAGES TAKEN BY THE VIKING 1 ORBITER IN 1980, DURING THE SOUTHERN WINTER. THE RED AREA ABOVE IS THE ARABIA REGION.

6 Where to Look for Signs of Life?

The astronauts of the Mars 1 expedition are, above all else, looking for life on Mars, present or past. Their work is the result of numerous activities undertaken since the time of the *Viking* spacecraft in 1976, including in situ analysis conducted by fixed or mobile automatic spacecraft, studies conducted on meteorites of Martian origin and research on soil and rock samples brought back to the Earth from the red planet. Studies have only given ambiguous results. No microorganisms, and no metabolism resembling that familiar to us on Earth, have been observed during these experiments, but minuscule objects with strange shapes have been uncovered, as well as poorly explained chemical activity. These results are not surprising. Scientists actually have no idea of the forms and manifestations of extraterrestrial life, appearing on a planet very different than the Earth and having survived in a much more severe environment.

The most widely held opinion in the scientific community is that life appeared on Mars in the same period as on Earth, during the first hundreds of millions of years after the formation of the solar system, when enormous impacts by asteroids and comets brought energy and water to the surface of both planets. This period of intense bombardment ended about 3.8 billion years ago, at a time when Mars and the Earth were probably the site of intense volcanic activity, and had dense atmospheres and oceans. The same causes produced, perhaps, the same effects. Martian life might have developed within the oceans, leading to microorganisms resembling our primitive bacteria. But then the two planets would have developed in very different ways. On Mars, the oceans disappeared, the atmosphere thinned and the surface was subjected to ultraviolet rays from the Sun; water was no longer present except in traces in the atmosphere, ice in the polar caps and perhaps permafrost in the subsoil.How could life have survived, even developed, in such an environment? What form might it have taken? Where would it have hidden?

From the beginning of the study of Mars from space, scientists shared the same idea: It would be necessary to look for actual life or fossils in regions where water seemed to have flowed or, better, pooled during long periods in the past. Special attention would need to be given to areas where water might still exist in a liquid state. Is this hypothesis reasonable? Would it be possible that Martian life simply adapted to the Martian environment and took forms that we couldn't even imagine – existing, for example, inside rocks or deep in the subsoil? Hasn't life on Earth taught us the extent to which microorganisms are capable of subsisting in extraordinarily inhospitable conditions, such as hot volcanic springs with temperatures reaching 265°F (130°C) or in frozen lakes? The adaptability of life is fantastic, and we can await many surprises on Mars and perhaps elsewhere in the solar system. The astronauts of Mars 1 must not therefore have any preconceptions; they need to look everywhere to seek to identify things that could indicate the presence of life.

1 LAUNCHING A WEATHER BALLOON CARRYING INSTRUMENTS TO MEASURE THE SOIL HUMIDITY AND SEEK THE MOST INTERESTING POINTS FOR EXPLORATION.

2 LEAVING FOR A LONG-TERM RESEARCH EXPEDITION.

3 AN ASTRONAUT EXAMINES CRACKS IN MARTIAN ROCKS, WHICH COULD HAVE PROTECTED A LIFE FORM.

4 EXTRACTING SAMPLES DEEP DOWN WITH A VEHICLE DESIGNED TO TAKE CORE SAMPLES..

1
2
3
4

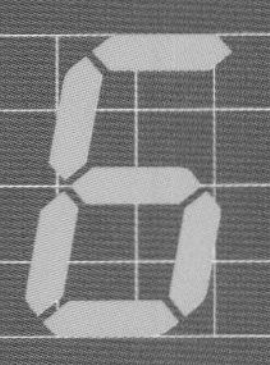

Martian Canyons

The two hemispheres of the red planet have very different appearances. North of the equator, Mars is mostly covered with large volcanic plains, formed after the initial episode of intense cosmic bombardment. These plains date from 2 billion to 3 billion years ago and were created by lava flows from the gigantic volcanoes in the Tharsis region. The Southern Hemisphere has a much more lunar appearance, with vast impact craters bearing witness to the force of the collisions with asteroids and comets during the first 700 million years of the solar system.

In the equatorial region between these two very different hemispheres, there is an extraordinary formation – a system of canyons extending over 2,500 miles (4,000 km) long and spreading 425 miles (700 km) wide at its central part. Some areas reach a depth of 4 miles (7 km). By comparison, the Grand Canyon in Arizona extends only 280 miles (450 km) and its maximum depth is only 1.2 miles (2 km). This collection of canyons was given the name Valles Marineris in honor of the *Mariner 9* spacecraft; it discovered this gigantic formation, which looked like a long scar on its images of the Martian globe. The origin of the Martian canyons is very different from that of the Grand Canyon dug out by the water of the Colorado River. Just to the west of Valles Marineris is the enormous bulge of the Tharsis region, which reaches 5.5 miles (9 km) in height and extends over 1,550 miles (2,500 km).

The uplifting of the ground, under the pressure of the magma that spread out from the three large volcanoes of Tharsis, probably caused cracks in the crust of Mars that became deep canyons. Other later phenomena – wind erosion and more or less violent flows of water – contributed to shaping them. The result is an area with a very complex appearance, with layers of sediment made visible in the cliffs, landslides, ravines, valleys reminiscent of dry washes in terrestrial deserts, and formations that speak of the shores of dried lakes or seas. Not all the canyons of Valles Marineris are connected together in the way a riverbed is connected to its tributaries. For example, in the north of the system, the Hebes Chasma depression is isolated from the valleys located farther south.

In its western part, where it meets the Tharsis bulge, Valles Marineris looks like a tangle of small canyons, called Noctis Labyrinthus. Farther to the east, two large parallel canyons break off – Tithonium Chasma to the north and Ius Chasma to the south. They continue into the largest part of Valles Marineris, with three vast parallel canyons: Ophir Chasma, Candor Chasma and Melas Chasma, which each have a length of around 125 miles (200 km).

In the east, all of the canyons converge in Coprates Chasma, which empties out on chaotic terrain, itself the source of valleys carved by violent flows of water and mud.

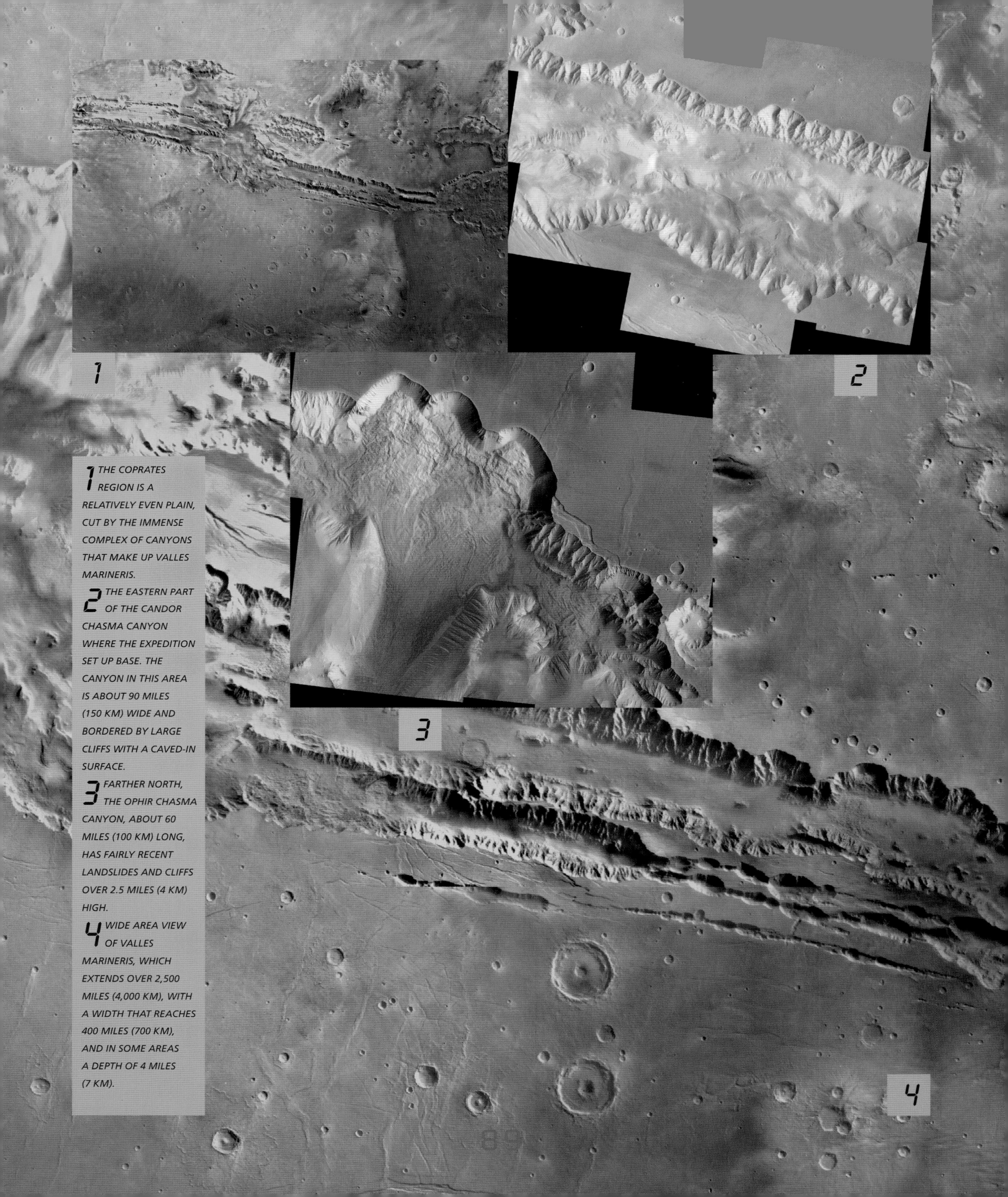

1 THE COPRATES REGION IS A RELATIVELY EVEN PLAIN, CUT BY THE IMMENSE COMPLEX OF CANYONS THAT MAKE UP VALLES MARINERIS.

2 THE EASTERN PART OF THE CANDOR CHASMA CANYON WHERE THE EXPEDITION SET UP BASE. THE CANYON IN THIS AREA IS ABOUT 90 MILES (150 KM) WIDE AND BORDERED BY LARGE CLIFFS WITH A CAVED-IN SURFACE.

3 FARTHER NORTH, THE OPHIR CHASMA CANYON, ABOUT 60 MILES (100 KM) LONG, HAS FAIRLY RECENT LANDSLIDES AND CLIFFS OVER 2.5 MILES (4 KM) HIGH.

4 WIDE AREA VIEW OF VALLES MARINERIS, WHICH EXTENDS OVER 2,500 MILES (4,000 KM), WITH A WIDTH THAT REACHES 400 MILES (700 KM), AND IN SOME AREAS A DEPTH OF 4 MILES (7 KM).

Valles Marineris

Candor Chasma, in the central part of Valles Marineris, is the site of the von Braun base. From there the astronauts can move out in the valleys and nearby plains, which offer an extraordinary variety of geological sites with connections to the history of the planet and, perhaps, Martian life.

The first objective, accessible with the Martian buggies, is the foot of the cliffs a few miles away from the base. The astronauts will study the mass of fallen rocks and try to understand the structure and origin of piled up layers several miles thick uncovered by the landslides.

They're also going to climb onto the plateau that separates Candor Chasma and Ophir Chasma and on which valleys that could have been carved by recent flows of water have been photographed.

Two long expeditions lasting a month, taken in the Martian motor homes, will be used to explore Valles Marineris towards the west, all the way to Noctis Labyrinthus, and to the east to the chaotic terrain where Coprates Chasma ends.

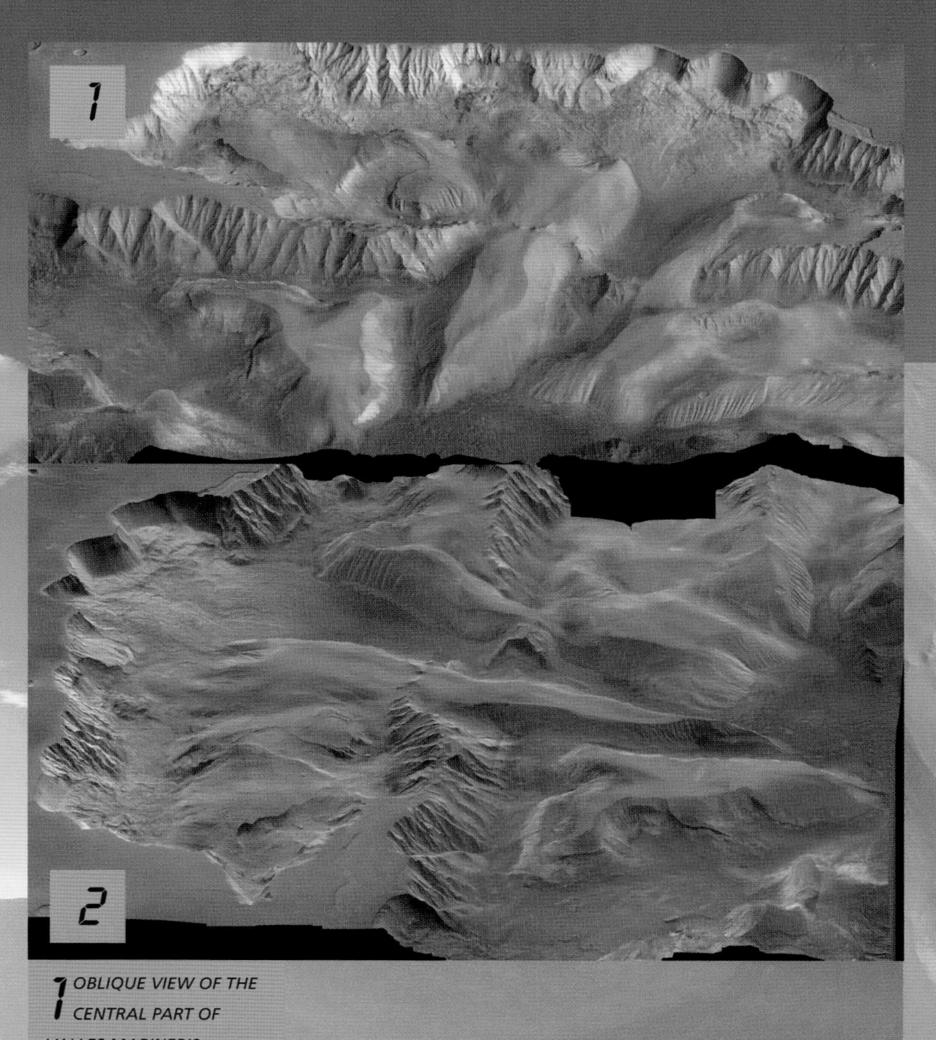

1 OBLIQUE VIEW OF THE CENTRAL PART OF VALLES MARINERIS, SHOWING AN INTERCONNECTION OF MULTIPLE CANYONS WITH PARTIALLY COLLAPSED CLIFFS.

2 CANDOR CHASMA (ON THE RIGHT) AND OPHIR CHASMA (ON THE LEFT), ARE UP TO 185 MILES (300 KM) WIDE IN THIS AREA. LAYERED DEPOSITS ARE VISIBLE AT THE BOTTOM OF CANDOR CHASMA ...

3 ... AND ARE ALSO VISIBLE IN THIS OTHER VIEW. THEY COULD HAVE BEEN CREATED BY AN IMMENSE LAKE THAT HAS NOW DISAPPEARED.

4 THE CLIFFS OF THE MARTIAN CANYONS ARE ESPECIALLY GOOD SITES TO STUDY, BUT THEIR EXPLORATION REQUIRES CLIMBING SKILLS.

6 The Volcanic Region

After the Valles Marineris canyons – long, tortured scars running down the face of the red planet – the most spectacular formations are volcanoes, the immense lava cones accumulated over many eruptions. Next to them, the ones we have on Earth are dwarfs. Volcanoes on Earth and Mars have the same origins. During the aggregation of the planet 4.5 billion years ago, and then through decay of radioactive elements, heat was released inside the planets and melted the rocks, creating a hot magma that rose to the surface. There, where the magma found a path through which to escape, volcanoes appeared. On Earth, a thin crust floats on the magma; this crust is cut in multiple pieces that are perpetually renewed: these are the tectonic plates. Hot points, where the magma rises from the depths, creating volcanoes, do not stay very long under the same zone of the Earth's crust, so volcanoes grow modestly. On Mars, in contrast, the crust is thick and fixed and the lava can flow through the very same openings for dozens or hundreds of millions of years.

1

Therefore, there are few volcanoes on Mars, but they are enormous.

The most remarkable region is the Tharsis bulge, to the northwest of Valles Marineris. It is an immense plateau rising 5.5 miles (9 km) above the planet's average surface and covers more than 11.5 million square miles (30 million sq km). Undoubtedly, this bulge resulted from the lift of an enormous volume of magma, part of which managed to break through the crust and flow out through volcanic mouths. Most visible are three giant calderas, which have a diameter of about 250 miles (400 km) and rise about 9 miles (15 km) above the Tharsis plateau; they are called Ascraeus Mons, Pavonis Mons and Arsia Mons.

These three volcanoes, practically aligned, are relatively young – just a few hundred million years old – with episodes of eruption perhaps not going back more than a few million years. Are they proof that Mars is still a hot and active planet, still the site of volcanic activity? It's up to us to find out. Pavonis Mons is the target of one of the big excursions in the *Gagarin* vehicle. The excursion will travel 1,100 miles (1,800 km) along Valles Marineris, then across Noctis Labyrinthus to reach the northeast flank of the volcano. The expedition's two astronauts, John Sturgett and Michel Morey, will collect rocks whose ages will be measured with precision. But they're especially going to seek to confirm an exciting hypothesis: About 10 million years ago, did rising magma cause an icy underground terrain to melt in this region, producing a gigantic flow of water toward the low northern plains of Mars? Could it be possible that water in a liquid state might continue to be present at shallow depths?

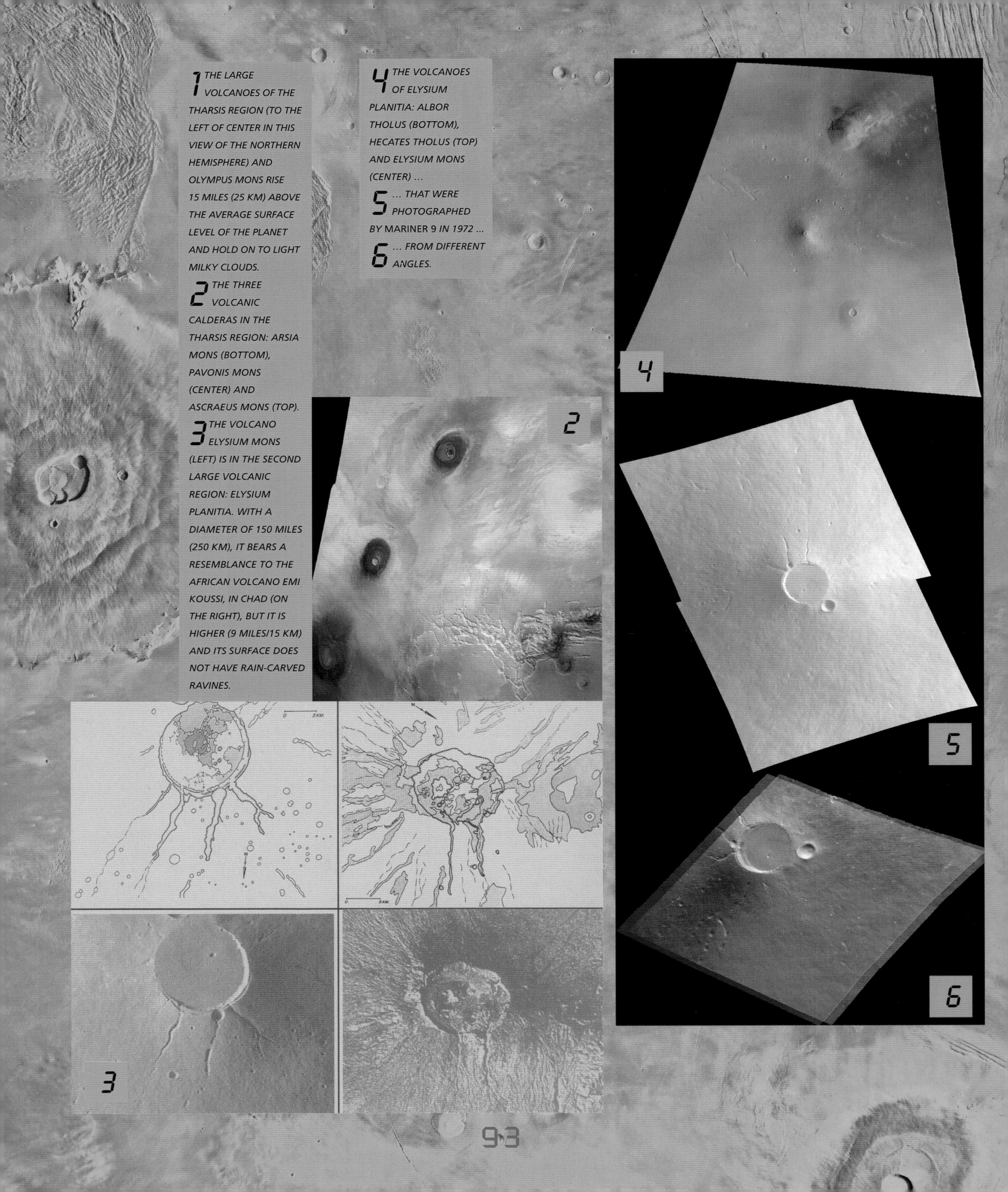

1 THE LARGE VOLCANOES OF THE THARSIS REGION (TO THE LEFT OF CENTER IN THIS VIEW OF THE NORTHERN HEMISPHERE) AND OLYMPUS MONS RISE 15 MILES (25 KM) ABOVE THE AVERAGE SURFACE LEVEL OF THE PLANET AND HOLD ON TO LIGHT MILKY CLOUDS.

2 THE THREE VOLCANIC CALDERAS IN THE THARSIS REGION: ARSIA MONS (BOTTOM), PAVONIS MONS (CENTER) AND ASCRAEUS MONS (TOP).

3 THE VOLCANO ELYSIUM MONS (LEFT) IS IN THE SECOND LARGE VOLCANIC REGION: ELYSIUM PLANITIA. WITH A DIAMETER OF 150 MILES (250 KM), IT BEARS A RESEMBLANCE TO THE AFRICAN VOLCANO EMI KOUSSI, IN CHAD (ON THE RIGHT), BUT IT IS HIGHER (9 MILES/15 KM) AND ITS SURFACE DOES NOT HAVE RAIN-CARVED RAVINES.

4 THE VOLCANOES OF ELYSIUM PLANITIA: ALBOR THOLUS (BOTTOM), HECATES THOLUS (TOP) AND ELYSIUM MONS (CENTER) ...

5 ... THAT WERE PHOTOGRAPHED BY MARINER 9 IN 1972 ...

6 ... FROM DIFFERENT ANGLES.

The Largest Volcano in the Solar System

Between the high volcanic plateau of Tharsis to the east, and the Amazonis Planitia to the west, the highest summit on the red planet, Olympus Mons, stands alone. It appeared on the old Martian maps, before the space era, under the name of Nix Olympica – the snows of Olympus – without knowing the nature of this light-colored spot on the face of Mars. Olympus Mons rises 15 miles (25 km) above the surrounding plain, and nearly 18 miles (30 km) above the lowest point on the red planet in Hellas Planitia. The highest summit on Earth, Mount Everest, stands less than 12 miles (20 km) above the deepest oceanic trench, east of Japan.

The gigantic Olympus Mons looks even more impressive when compared to the largest terrestrial formation of the same kind: Mauna Kea, a shield volcano that belongs to the Hawaiian Islands in the Pacific. Mauna Kea rises only 5 miles (9 km) above the ocean bottom, and its diameter is only 75 miles (120 km), as compared with 375 miles (600 km) for Olympus Mons! The difference is even more striking if the two volumes are compared: Olympus Mons has a volume six hundred times that of Mauna Kea. It bears witness to colossal eruptions in a past that is not that distant – within 100 million years.

Undoubtedly less interesting to science than Pavonis Mons, its symbolic value is nonetheless too significant for the Mars 1 expedition to leave it off its exploration program. Too far from the von Braun base for surface expedition, Olympus Mons is the objective of the first trip on board the Martian shuttle *Red Challenger*, named in honor of the NASA shuttle destroyed in a catastrophic explosion in 1986. Otto Kruger and Nagatomo Itochu are the crew. *Red Challenger* is a rocket glider with very large and thin wings, and almost looks like a butterfly; it is at the edge of material science in the 2030s. It takes off like a rocket, rises a few dozen miles in altitude and glides toward its objective. On the way to Olympus Mons, it flies over Noctis Labyrinthus and the Tharsis plateau, offering magnificent perspectives. The view from Olympus Mons is a little surprising: the average slope of its sides is low – less than 5 degrees – and the volcano appears to be nearly flat; even the 4-mile (6 km) high cliff that surrounds it is slightly inclined. After its flight of only half an hour, *Red Challenger* lands vertically on the edge of the great caldera – the collapsed lava lake that occupies the summit of the volcano and measures 50 miles (80 km) in diameter. Humanity has reached the highest summit in the solar system.

1 OLYMPUS MONS IS A VOLCANO 375 MILES (600 KM) IN DIAMETER ...

2 ... WHICH RISES MAJESTICALLY ABOVE THE SURROUNDING PLAINS, FROM WHICH IT IS SEPARATED BY A 4-MILE (6 KM) HIGH CLIFF.

3 AERIAL VIEW FROM ABOVE VALLES MARINERIS WITH THE MARTIAN VOLCANOES IN THE THARSIS REGION IN THE DISTANCE ...

4 ... AND, ON THE HORIZON, THE SILHOUETTE OF OLYMPUS MONS.

5 THE SUMMIT OF OLYMPUS MONS IS COVERED BY A SERIES OF NESTED CALDERAS – COLLAPSED LAVA LAKES FORMED DURING SUCCESSIVE ERUPTIONS.

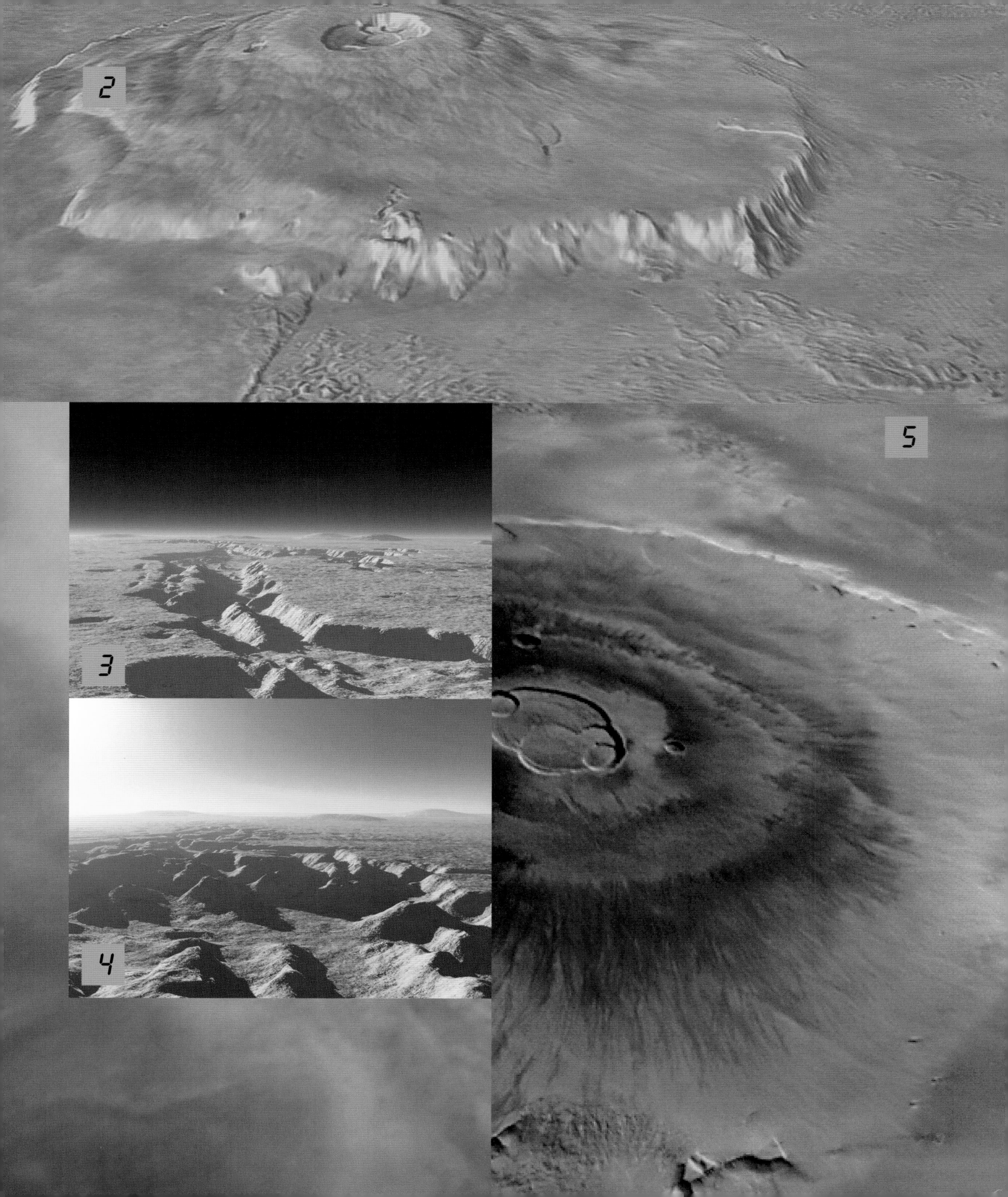
2
3
4
5

Martian Craters

During the first hundreds of millions of years of its existence, the solar system was the site of gigantic collisions between planets, their satellites and thousands of asteroids and comets that had not yet found stable orbits in certain zones, like the asteroid belt between Mars and Jupiter, the Kuiper belt beyond Neptune or the Oort cloud even farther out. These gigantic collisions created immense depressions, enormous craters, which indelibly marked the surface of many objects – the Moon, Mercury and Mars. On Mars the large basins in the southern hemisphere, like Argyre and Hellas, are the scars left by these titanic, cosmic confrontations. Later the solar system became calmer, but cosmic collisions did not stop – some smaller asteroids and more modest comets continued in the bombardment. The pockmarked faces of the Moon and Mercury bear witness to these billions of years of cataclysms. Earth was not spared: a collision with a comet likely explains the disappearance of the dinosaurs 65 million years ago. But craters made by those ancient impacts have been wiped away by erosion and renewal of the Earth's crust.

On Mars, the situation is in between that of the Moon and Earth. The great basins and large craters remain, even if their contours are partially destroyed by volcanic eruptions as well as wind erosion and, maybe, even water erosion. Their bottoms are covered with sediment of various origins. The smaller craters only remain in the regions where the surface has not been covered over by dust or sediments. The density of craters is therefore an indication of the age of the surface; a lunar appearance indicates a very ancient surface, while a surface with little marking from impacts is certainly young. The study of crater density reveals astonishing differences. The old landscapes are principally located in the Southern Hemisphere, which on average is much less elevated than the northern half of the planet. In contrast the North Pole is surrounded by an immense depression, largely free of craters. What is the cause of this asymmetry? Were there more enormous asteroid impacts that dug immense basins that are today masked by volcanic outflows and ... sediment from a vast northern ocean? To begin to get an answer, the *Red Challenger* shuttle has been sent toward the Galle Crater on the edge of one of the large Southern Hemisphere basins: Argyre Planitia. Its results will be compared to those from studies conducted in the region of the possible northern ocean.

1

1 THE GALLE CRATER MEASURES 135 MILES (215 KM) IN DIAMETER AND IS LOCATED IN THE HIGH SOUTHERN LATITUDES ...

2 ... LIKE THE LOMONOSOV CRATER, COVERED WITH FROST IN WINTER.

2

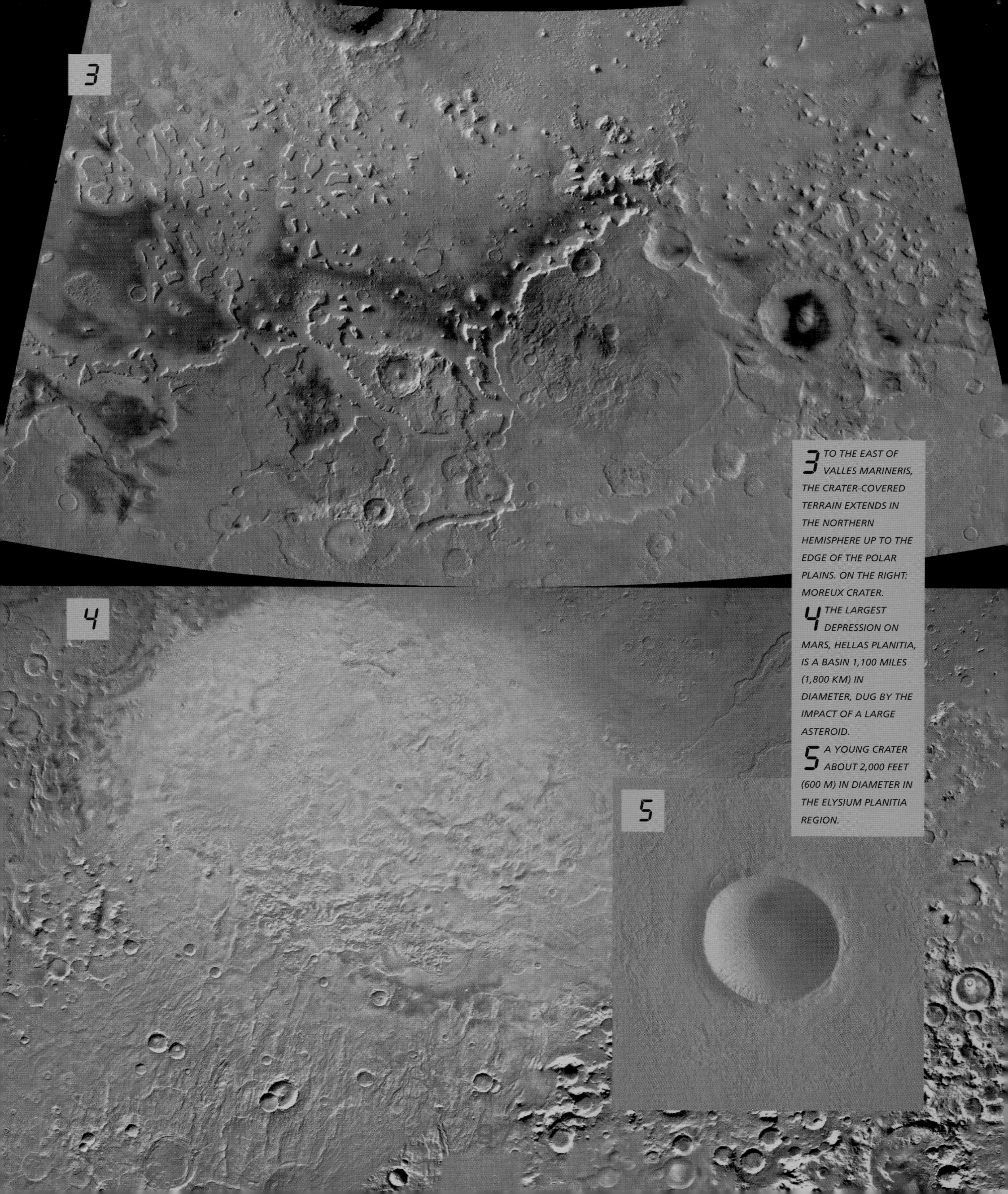

3 TO THE EAST OF VALLES MARINERIS, THE CRATER-COVERED TERRAIN EXTENDS IN THE NORTHERN HEMISPHERE UP TO THE EDGE OF THE POLAR PLAINS. ON THE RIGHT: MOREUX CRATER.

4 THE LARGEST DEPRESSION ON MARS, HELLAS PLANITIA, IS A BASIN 1,100 MILES (1,800 KM) IN DIAMETER, DUG BY THE IMPACT OF A LARGE ASTEROID.

5 A YOUNG CRATER ABOUT 2,000 FEET (600 M) IN DIAMETER IN THE ELYSIUM PLANITIA REGION.

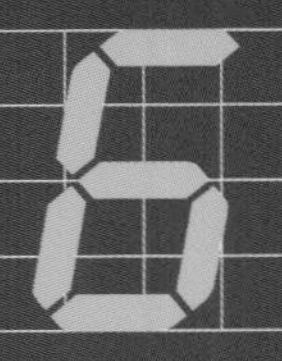

Dried Rivers and Lakes

Unlike the Grand Canyon on Earth, the grand canyons of Valles Marineris where the von Braun base was set up were not dug by running water. But other formations irresistibly evoke the action of rivers and streams, even the prolonged presence of water in lakes or seas. How is that possible if the red planet today is a hopelessly cold and dry desert? Are we to suppose that a valley like Nirgal Vallis, at the southeast of Valles Marineris, apparently joined by a network of tributaries, was carved by a river during a period when a more temperate climate existed on Mars? What other explanations could account for these familiar forms? It is true that Nirgas Vallis and the other valleys observed are different from those encountered on Earth; their sources are not located on heights cut with ravines by rain, but on plateaus free of erosion, and they often stop in the middle of nowhere, in regions that do not resemble dried lakes or seas. Could they have formed by the flow of very fluid lava? By subsurface rivers whose covering caved in? Could carbon dioxide gas have played a role?

The most likely hypothesis is liquid water on the surface or subsurface. To verify it, a large expedition will be dedicated to the exploration of Nirgal Vallis, onboard *Glenn*, whose resources will be stretched to the maximum, as the distance to travel is over 1,200 miles (2,000 km).

In parallel, another expedition, with the other vehicle, *Gagarin*, will visit other valleys, this time to the northeast of Valles Marineris. There, Tiu Vallis and Ares Vallis, which could have been formed by cataclysmic floods, undoubtedly resemble – although much more powerful – the violent rivers that arise from rainstorms and dig the wadis in the Maghreb. The presence of the second type of valley, which often starts in chaotic terrain and ends in depressions, especially in the large northern plains, could be explained if it is assumed that the Martian subsoil contains (or once contained) a mixture of soil and ice, like Siberian permafrost. A massive melting of the permafrost, resulting, for example, from a rise of magma, would be able to cause a sudden and colossal flood running toward low-lying areas nearby. These valleys are immense: more than 600 miles (1,000 km) long, 60 miles (100 km) wide and several miles deep. They bring to mind Dantesque floods, dozens of times more intense than the Amazon, perhaps even thousands of times more intense! Some formations observed to the northwest of Arsia Mons, one of the volcanoes in the Tharsis dome, could be interpreted as having been created by floods 50,000 times stronger than the Amazon. In a few weeks, flows like that could fill a body of water comparable to the Indian Ocean and cause an extreme climate change. Does that mean that the Martian climate had significant variations, with a sudden appearance of temperate episodes? The expeditions in the northern plains will seek to test these hypotheses, and in particular the possible former existence of an ocean around the North Pole.

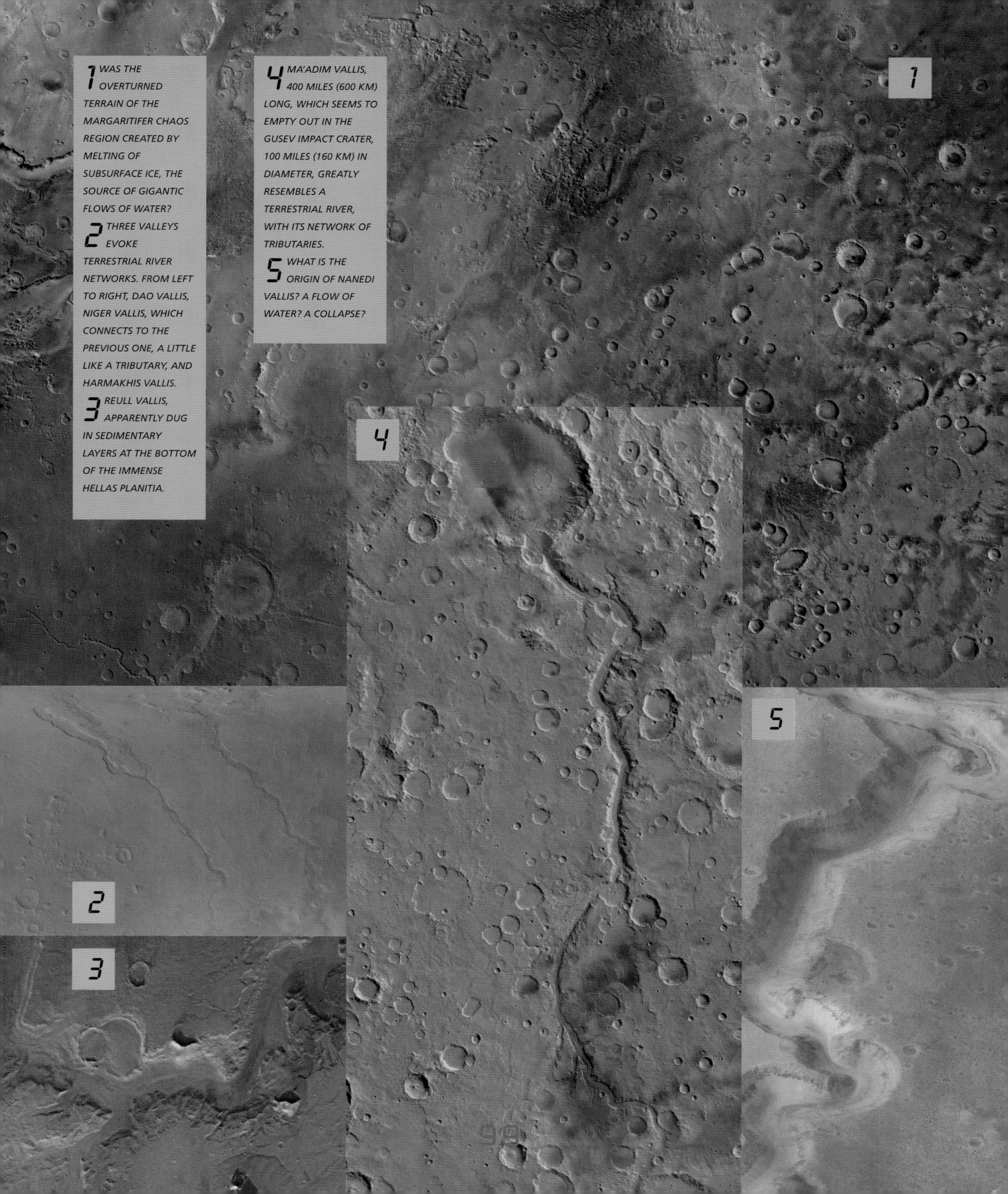

1 WAS THE OVERTURNED TERRAIN OF THE MARGARITIFER CHAOS REGION CREATED BY MELTING OF SUBSURFACE ICE, THE SOURCE OF GIGANTIC FLOWS OF WATER?

2 THREE VALLEYS EVOKE TERRESTRIAL RIVER NETWORKS. FROM LEFT TO RIGHT, DAO VALLIS, NIGER VALLIS, WHICH CONNECTS TO THE PREVIOUS ONE, A LITTLE LIKE A TRIBUTARY, AND HARMAKHIS VALLIS.

3 REULL VALLIS, APPARENTLY DUG IN SEDIMENTARY LAYERS AT THE BOTTOM OF THE IMMENSE HELLAS PLANITIA.

4 MA'ADIM VALLIS, 400 MILES (600 KM) LONG, WHICH SEEMS TO EMPTY OUT IN THE GUSEV IMPACT CRATER, 100 MILES (160 KM) IN DIAMETER, GREATLY RESEMBLES A TERRESTRIAL RIVER, WITH ITS NETWORK OF TRIBUTARIES.

5 WHAT IS THE ORIGIN OF NANEDI VALLIS? A FLOW OF WATER? A COLLAPSE?

ON MARS

6 The Mystery of Martian Water

Was there lots of water on Mars billions of years ago, with lakes filling large craters dug by asteroids impacts, and maybe even a vast northern ocean around the North Pole? This is not impossible. The gases given off by the eruptions of the enormous volcanoes in the Tharsis dome could have surrounded the red planet with an atmosphere denser than the Earth has today and created, through the greenhouse effect, a climate warmer and more humid than the present one. In these remote periods, Mars could even have possessed, in proportion to its mass, more water than the Earth. But when did these aquatic surfaces disappear? Did most of the water escape into space, like the planet's original atmosphere? Does a significant amount of ice remain in the form of permafrost as the observations of *Mars Odyssey* seemed to show at the beginning of the 21st century? Did lakes, even oceans, sometimes re-form during the course of drastic climate changes brought about by volcanic or astronomical phenomena? (Mars' axis of rotation can change direction significantly, unlike that of the Earth which benefits from the stabilizing effect of the Moon.) If there were any changes, how long did these episodes of temperate climate last? When was the last episode?

Many of these questions should find at least partial answers with the exploration of the Holden Crater, a formation 110 miles (175 km) in diameter, with thick sediments at its bottom. Were the sediments deposited by a lake? The team leaves to study Nirgal Vallis, onboard Glenn, and will also explore the crater located in the same area. The search for the past existence of lakes will be the subject of another expedition, with the *Red Challenger* shuttle. Its destination will be the Elysium Basin, south of the volcanic bulge of the same name on which arises Hecates Tholus, Elysium Mons and Albor Tholus. The Elysium Basin, with a width of 1,900 miles (3,000 km) could have been the site of a lake 5,000 feet (1,500 m) deep.

The existence of aquatic surfaces on Mars, in the more or less distant pasts, is a fundamental question. But another question is even more important: Is liquid water still present today near the surface? In some way, it involves looking for Martian oases, where sources of subsurface heat maintain favorable conditions for the potential survival of microorganisms. Ravines with very recent origins have been observed on the edges of many craters, mainly in the Southern Hemisphere. Some of the most spectacular are visible in the Newton Crater, a vast circle nearly 200 miles (300 km) in diameter, in the cratered highlands south of Daedalia Planum. To travel to this site far from the von Braun base, the *Red Challenger* shuttle will be needed.

1 IN CANDOR CHASMA, NOT FAR FROM THE VON BRAUN BASE. WERE THESE SEDIMENTS DEPOSITED AT THE BOTTOM OF LAKES?

2 RECENTLY FORMED RAVINES INSIDE NEWTON CRATER ...

3 ... AND IN ANOTHER CRATER IN THE NOACHIS TERRA REGION.

4 WHEN AND HOW WERE THESE STRIKING RAVINES FORMED?

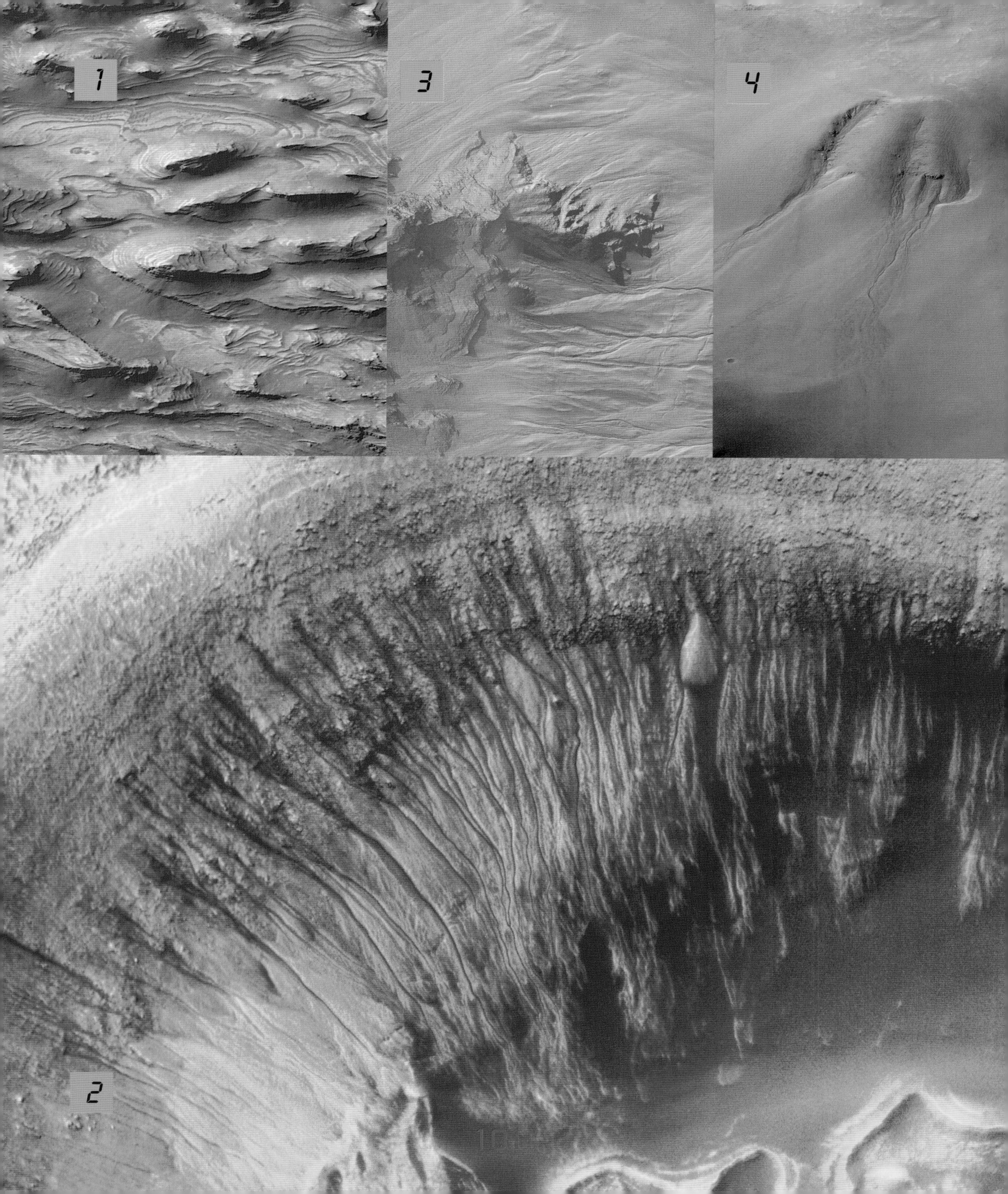
1
3
4
2

6 The Martian Poles

Where should we look for water on Mars? This question is at the heart of the exploration of what appear to be dry valleys and lakes. But there is a region where there is no doubt that water is present in large quantities: the permanent ice cap that covers the North Pole. The cap measures about 750 miles (1,200 km) in diameter and is over 2 miles (3 km) thick. It contains about 360,000 cubic miles (1.5 million km3) of ice. This quantity is small compared to the volume of the Antarctic ice on Earth, but still represents about half of the glaciers on Greenland. Melted, this frozen water would cover the surface of Mars with a layer 30 feet (10 m) deep. The shape of the red planet's northern polar cap is tortured, with curved canyons whose depth can reach over half a mile (1 km). Further, it has a laminated structure with layers dozens of feet thick piled up for miles. The surprising appearance is probably due to a succession of climate changes recorded in the successive layers of ice and dust. The exploration of this immense zone will be of interest to all the Martian expeditions to come, but Mars 1 cannot leave this region untouched. *Red Challenger* is going to land at the edge of the largest glacial canyon, Chasma Boreale, in order to start to read the history of the Martian climate on the edges of this deep valley. Could some form of life have survived there? The temperature is always very low: below –85°F (–65°C). But it's known that water very rich in salts can stay liquid at temperatures nearly as low.

In winter, the northern polar cap spreads considerably: Carbon dioxide gas is deposited in a solid state all the way to 60 degrees latitude. For the southern polar cap, its temperature (–165°F/–110°C in the summer) is sufficiently low for carbon dioxide gas to remain permanently frozen. It is therefore composed of a mixture of water and carbon dioxide ices. In the summer, its diameter does not extend more than 220 miles (350 km). This very significant disparity in composition and size of the northern and southern polar caps is related to the seasons, which are different in the two hemispheres. In the north, the summer is cold and long; in the south, the summer is hotter and shorter. The dust storms that arise during the southern summer and northern winter lead to deposits of dust layers at the North Pole, favoring the condensation of ice. The South Pole, much less rich in water, will therefore be explored later.

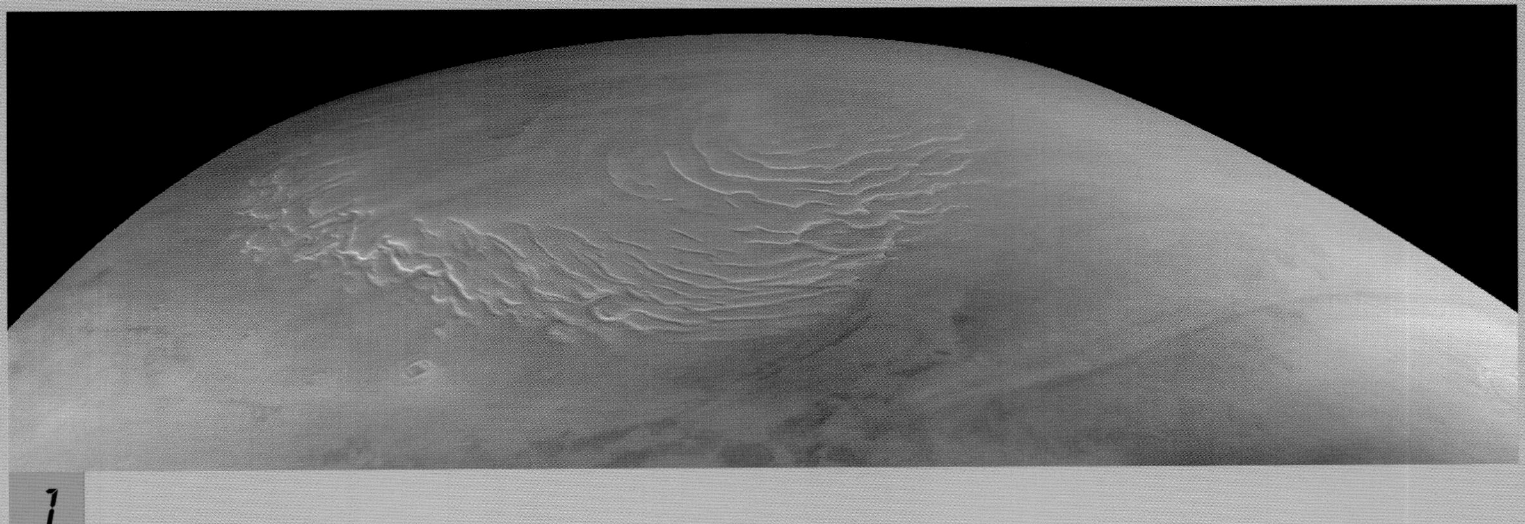

1

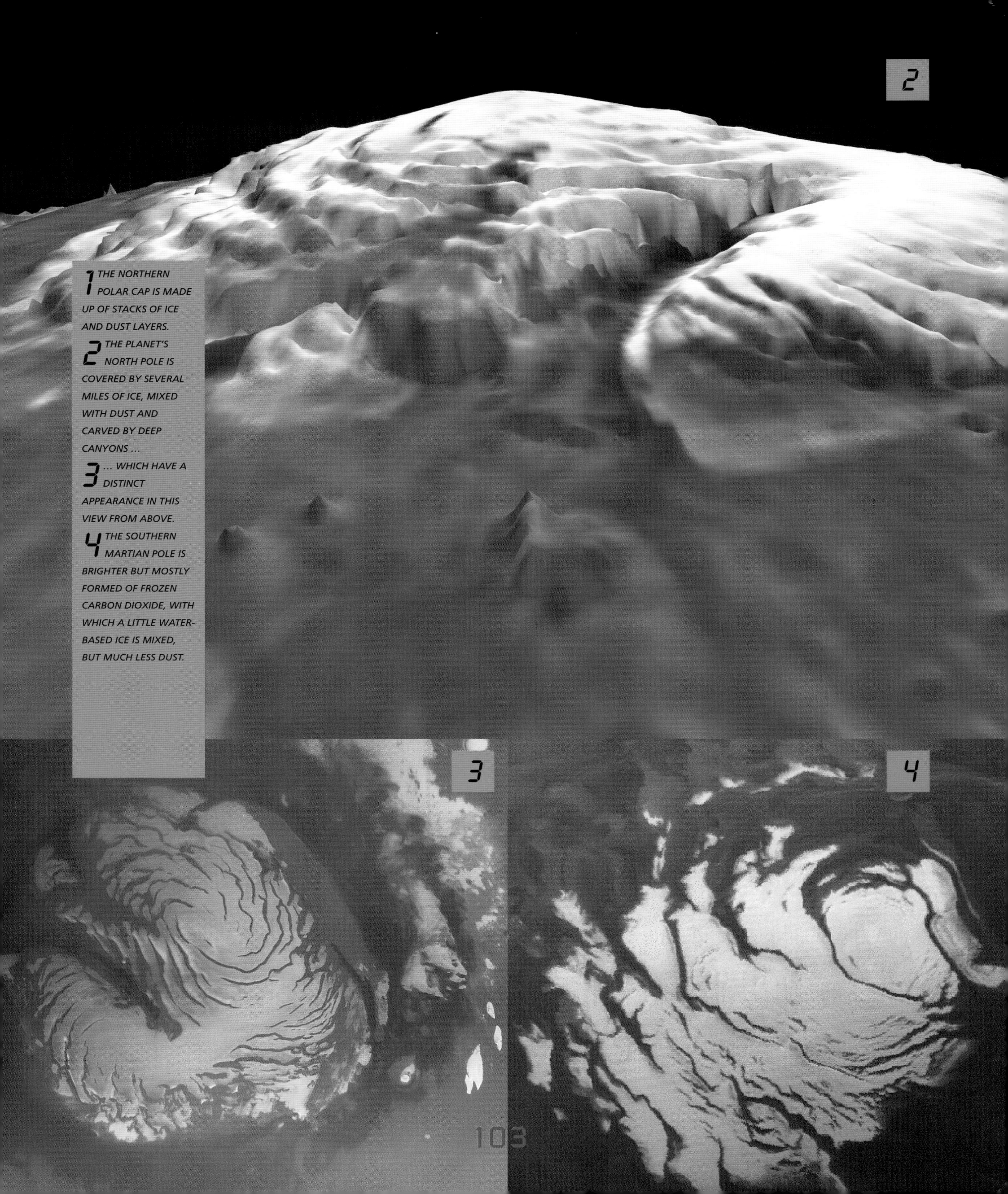

1 THE NORTHERN POLAR CAP IS MADE UP OF STACKS OF ICE AND DUST LAYERS.

2 THE PLANET'S NORTH POLE IS COVERED BY SEVERAL MILES OF ICE, MIXED WITH DUST AND CARVED BY DEEP CANYONS ...

3 ... WHICH HAVE A DISTINCT APPEARANCE IN THIS VIEW FROM ABOVE.

4 THE SOUTHERN MARTIAN POLE IS BRIGHTER BUT MOSTLY FORMED OF FROZEN CARBON DIOXIDE, WITH WHICH A LITTLE WATER-BASED ICE IS MIXED, BUT MUCH LESS DUST.

Drilling on Mars

There is no room for doubt that there is no liquid water and there are no living microorganisms on the surface of Mars today. The physical conditions – low pressure, very low temperature, ultraviolet radiation from the Sun and so on – do not allow it. The study of the Martian soil, in regions as varied as the large canyons, volcanoes, basins, craters, valleys and polar caps teach us much about the rocks, their age and their geological and climatic history. The search for the presence of life only makes sense in the depths. For all the robots sent to remote parts of the planet and all the expeditions conducted in vehicles or shuttles that drilled for soil samples, their means were limited and did not allow taking core samples from more than several dozen feet below the surface. The only exception was at the northern polar cap, where cryonic robots (those that melt the ice to drill down in it) were able to go several hundred feet in the subsoil.

At the von Braun base, a system directly inspired by oil drilling equipment can obtain samples from several miles deep. Unfortunately, this equipment is heavy and impossible to transport long distances from the base to places where there could potentially be outcrops of subsurface ice, and where liquid water might exist near the surface. It is therefore necessary to

be satisfied with drilling in Candor Chasma, with the hope of bringing back some proof of the existence of permafrost under most of the Martian surface. If it does exist, how deep is it? The estimates range from about 1 to 2 miles (2 to 3 km) in the Northern Hemisphere, and 3 to 4 miles (5 to 6 km) in the Southern Hemisphere (although *Mars Odyssey* might've left the hope that the "dirty ice" could be very close to the surface in certain regions, notably near the South Pole). Could there be a life form at these depths? It's not impossible. On Earth, bacteria have been discovered a few miles down, in environments without oxygen; however, those environments are hot. What will be found in the core samples that are being drawn from the depths of the red planet?

ON MARS

6 Rally in the Martian Dunes

Despite the tenuous nature of the atmosphere, the wind is a major factor in shaping the Martian landscape. It lifts and transports the very fine dust, which covers nearly the entire surface of the planet, over large distances. The great storms that arise in the Southern Hemisphere during the summer, at the time of perihelion opposition, completely change the appearance of some regions. Undoubtedly this explains the different maps of the red planet traced by careful observers in the 19th century during different oppositions. Even though, when protected by their spacesuits, the astronauts hardly feel the Martian wind, it is capable of causing a significant erosion of the relief over long periods. In the low areas of Vastitas Borealis near the North Pole, thousands of hills, mesas and other typical formations demonstrate the wind's work. One of the hills eroded by the wind has even become celebrated in a photo taken by the *Viking* spacecraft. Its shape resembled a human face, and the resemblance was sufficiently striking that several articles and a book were dedicated to it, explaining that it had been a monument constructed by a Martian civilization. The dreams of Percival Lowell just won't die. This formation even has a key role in Brian De Palma's film, *Mission to Mars*. Clearly, it doesn't involve anything more than an optical illusion arising from the limited precision of the *Viking* images. *Mars Global Surveyor* established the truth; perhaps that's too bad.

The combination of dust grains and storms have an obvious effect – the formation of dunes, especially in the depressions where the dust can accumulate. There are also immense dune fields with varied and surprising shapes, both in the planet's northern and southern plains and in large craters. Unless they're specifically researching them, coming across them gives the crew members of the *Tsiolkovski* an opportunity to use their buggy in an unusual manner. The crossings are more sport than science, resembling the famous trans-Saharan rallies on Earth. They even organized a race in the dune fields near the von Braun base.

1

1 THIS HILL SCULPTED BY MARTIAN WINDS SOMEWHAT RESEMBLES A HUMAN FACE. IT IS THE "FACE OF MARS" WHICH WOKE UP DREAMS FROM THE TIME OF PERCIVAL LOWELL.

2 DUNE FIELDS IN NILI PATERA, A DEPRESSION IN THE CENTRAL PART OF SYRTIS MAJOR.

3 DUNES AT THE BOTTOM OF ONE OF THE LARGE MARTIAN CRATERS, HERSCHEL, 200 MILES (320 KM) WIDE.

4 ON THE RED PLANET THE DUNES CAN TAKE CLASSICAL FORMS ...

5 ... OR, ON THE OTHER HAND, VERY STRANGE ONES.

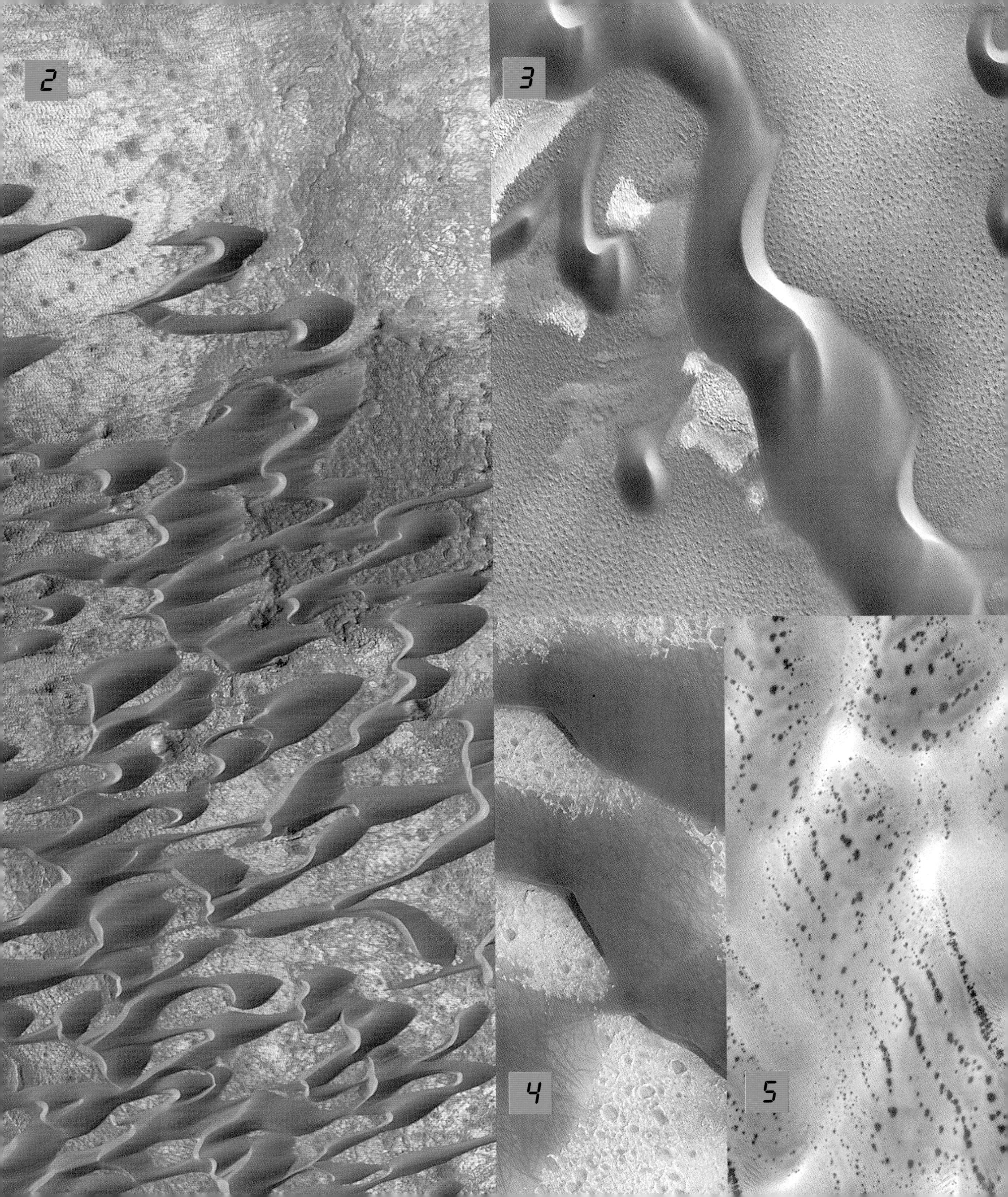
2
3
4
5

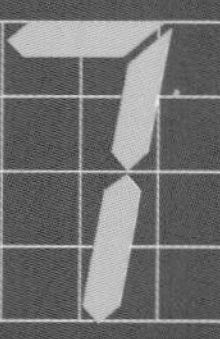

Leaving Mars

1 ON A FINAL EXPEDITION'S RETURN TO CHRYSE PLANITIA, AN ASTRONAUT MAKES A SYMBOLIC VISIT TO THE VIKING 1 SPACECRAFT STANDING THERE ON THE REGION'S ROCKY SOIL SINCE 1976.

2 AUGUST 4, 2035: TAKEOFF FOR PHOBOS... IS IT FAREWELL TO MARS OR SEE YOU LATER?

More than 20 months on Mars ... that can seem long, more so because as time passes the contacts with Earth, both professional and personal, have more and more often been maintained through e-mail, with pictures or movies of course, but without real-time exchanges. Signals that take up to 20 minutes to make a round trip between the parties do not allow a real-time conversation. Further, from day to day, the difference between Martian and Earth times shifts little by little and they quickly forget what could be a good time of day there on Earth, the small bluish point sometimes visible in the nighttime sky. The planners of the Mars 1 expedition hadn't expected this: There is a time on Mars and a time on Earth, during which real-time contact is lost very quickly. An interplanetary expedition is essentially a unique experience of solitude shared by a few people, in a foreign world whose unique beauty further isolates them from colleagues, friends and family left behind on Earth. In some fashion, Mars possesses the astronauts; if it were possible, some of them would stay there without any hesitation. The planet has changed their life. Nothing will ever be as it was before.

The colonization of Mars is not yet a reality. They have to return, and that means loading into the Mars-Phobos shuttle, which will take them away in a few weeks, along with all the samples of soil and rocks, core samples brought up from the depths, photographs, video recordings, preliminary analyses, expedition reports and so on. That's several tons of scientific data that the researchers in Earth-based laboratories await with immense interest.

Some of this material is actually already aboard the *Tsiolkovski*, which is waiting near Phobos. Nagatomo Itochu and Otto Kruger have made two short visits to the Phobos base to check on the production of hydrogen necessary for

the return trip, and to check on the *Tsiolkovski*. Many results have already been transmitted by e-mail, but their content has actually increased the impatience of scientists on Earth. The final task is to put the von Braun base in standby mode, ready to welcome future crews. Then, August 4, 2035, is the time for the departure.

7 The Stages of Return

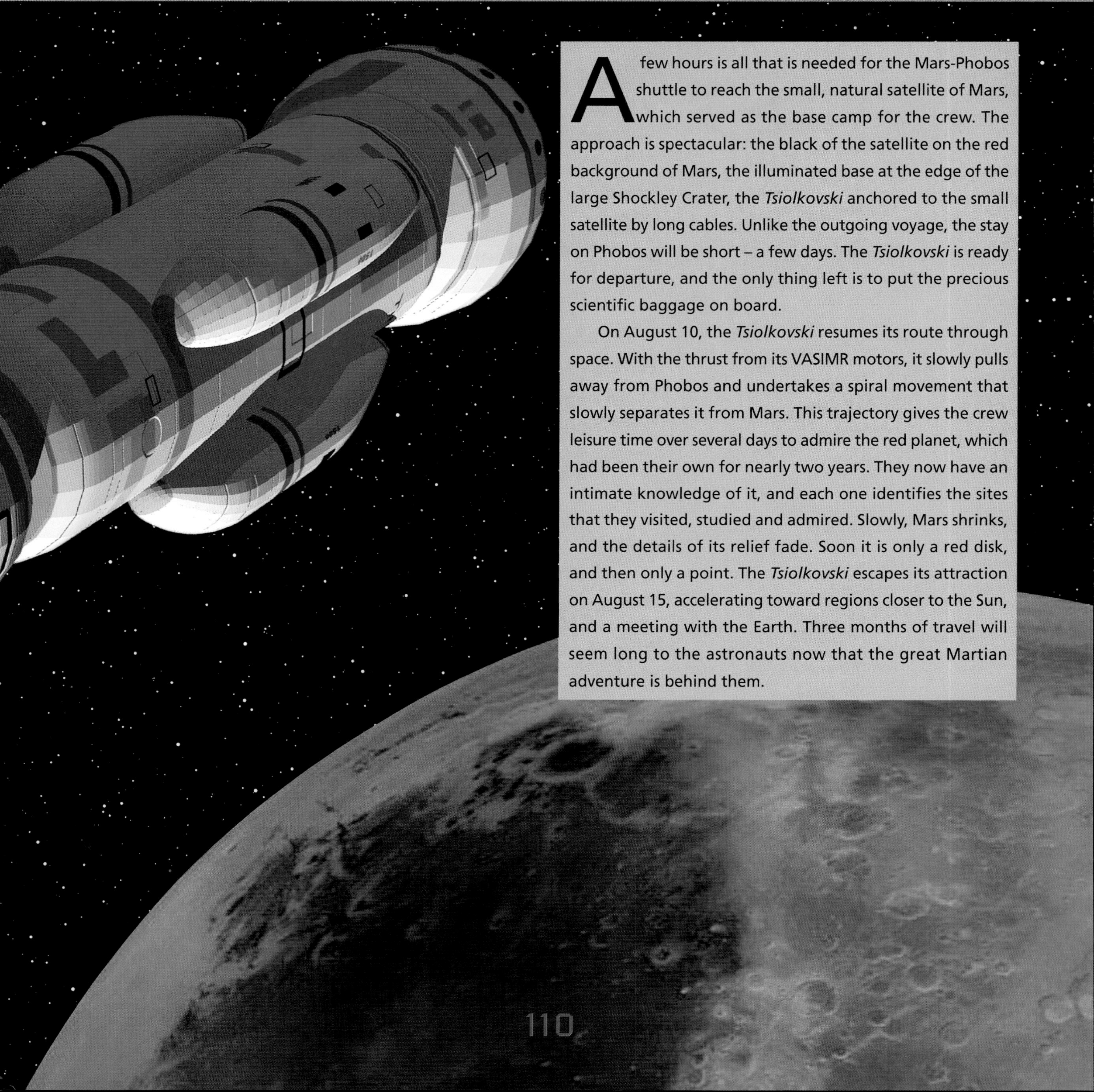

A few hours is all that is needed for the Mars-Phobos shuttle to reach the small, natural satellite of Mars, which served as the base camp for the crew. The approach is spectacular: the black of the satellite on the red background of Mars, the illuminated base at the edge of the large Shockley Crater, the *Tsiolkovski* anchored to the small satellite by long cables. Unlike the outgoing voyage, the stay on Phobos will be short – a few days. The *Tsiolkovski* is ready for departure, and the only thing left is to put the precious scientific baggage on board.

On August 10, the *Tsiolkovski* resumes its route through space. With the thrust from its VASIMR motors, it slowly pulls away from Phobos and undertakes a spiral movement that slowly separates it from Mars. This trajectory gives the crew leisure time over several days to admire the red planet, which had been their own for nearly two years. They now have an intimate knowledge of it, and each one identifies the sites that they visited, studied and admired. Slowly, Mars shrinks, and the details of its relief fade. Soon it is only a red disk, and then only a point. The *Tsiolkovski* escapes its attraction on August 15, accelerating toward regions closer to the Sun, and a meeting with the Earth. Three months of travel will seem long to the astronauts now that the great Martian adventure is behind them.

1 BEFORE RESUMING THE TRIP TO EARTH …

2 … A SHORT STOPOVER AT THE BASE ON PHOBOS WHERE THE TSIOLKOVSKI *IS ANCHORED.*

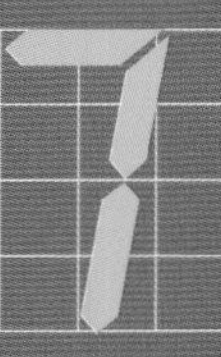

In Quarantine

Earth approaches and the astronauts can now start to distinguish the outline of the continents with the naked eye. The blue planet is a beautiful sight! Nostalgia grips them each time they think of Mars, but it's no longer the time to think about their experience there. They had plenty of time to do that over the last three months during their uneventful trip. On the approach to Earth, the *Tsiolkovski* will be met by an Earth-Moon shuttle, allowing the crew to get to the International Space Station (ISS) more quickly with the scientific materials that they are bringing back. There is no crew on the shuttle, as the astronauts are considered to be dangerous! Of course, none of them were infected or the victim of unexplained symptoms. All of them remained in excellent health on the red planet, far from terrestrial microbes. But it's better to be safe, and a quarantine of a few weeks is imposed, in a special module attached to the ISS. There they must rid themselves of all the clothing and objects from the expedition, which will be carefully studied and sterilized.

The scientific materials will be placed in an isolated Class 4 research laboratory for bacteriological safety. Class 4 is the highest level; it allows work with the most dangerous microorganisms, like the Ebola virus. After this first phase of study, some materials will be sterilized and sent to other Class 4 laboratories on Earth, like that in Lyon, France. During automated missions to recover samples, only sterilized samples from Mars were brought directly to Earth and sent to such laboratories. But this time it is impossible to sterilize all the scientific material since the astronauts are part of it! This is certainly an excellent idea from a scientific perspective, but the risk from it is heightened.

1 ONBOARD THE INTERNATIONAL SPACE STATION IN A HIGH SECURITY LABORATORY, THE STUDY OF THE SCIENTIFIC MATERIAL FROM MARS BEGINS ...

2 ... WHILE THE CREW FROM THE MARS 1 EXPEDITION STAYS IN QUARANTINE.

7 Is the Mystery Solved?

Did the Mars 1 expedition meet the hopes that were placed on it? Will it allow resolution of humanity's great questions about the appearance of life, its own evolution and its place in the universe? That is difficult to answer, because the research based on the samples brought back will take years, undoubtedly even decades. In 2002, three decades after the return of the last astronaut from the *Apollo* project, the entirety of the lunar soil samples gathered during that program had not yet been analyzed. And the quantity and variety of scientific material brought from the red planet by the Mars 1 astronauts is on an entirely different scale from the 853 pounds (387 kg) of lunar soil brought back by the *Apollo* spacecraft. But isn't there already the beginning of an answer to the only question that really counts for the public at large: Had a Martian life form been discovered?

The preliminary analysis revealed signs suggestive of microorganisms in most of the samples gathered inside rocks, or from deep samples, like those observed earlier in some meteorites, but with an incredible variety of shapes and sizes. Are these really fossil remains of Martian microorganisms? Is their variety linked to subsequent appearances of life forms on Mars, unlike the Earth where life, apparently, started only once? Do some of these life forms still exist in underground oases? These oases were not found, but that proves nothing. Seven astronauts and two years of exploration can seem like a lot, but is it when compared with the immensity of a planet?

The exploration of Mars is only beginning. It will be even more exciting than the first results confirming a past rich in violent episodes and spectacular climate changes. Unlike the Earth, whose climate has changed relatively little over the ages – perhaps through the stabilizing influence of the biosphere, with its plants and animals – Mars appears to be an unstable world. It is a site of cataclysms of an unimaginable magnitude: giant volcanic eruptions, monstrous cataracts emptying into the planet's basins and creating seas in a few weeks. Is that a lesson for people on Earth? Isn't the comfortable environment on the blue planet an exception, an incredible stroke of good luck? Shouldn't it be preserved at all costs? But this situation may open up extraordinary possibilities. If Mars is unstable, would it be possible to swing it into a much more temperate state, to make it an inhabitable planet, a second Earth? Some dream about it, but could it truly be done?

1 NEARLY TWO YEARS OF COLLECTING CORE SAMPLES, DRILLING AND EXPLORATION ALLOWED THE MARS 1 EXPEDITION TO BRING BACK THOUSANDS OF SAMPLES …

2 … WHILE UP TO NOW SCIENTISTS ONLY HAD A FEW METEORITES TORN FROM THE RED PLANET …

3 … AND SOME ROCKS BROUGHT BACK BY AUTOMATED SPACECRAFT.

4 IT WILL TAKE YEARS, MAYBE DECADES, TO TAKE ADVANTAGE OF ALL THIS MATERIAL …

5 … WHICH IS NOT SURPRISING, BECAUSE THE EXPLORATION OF THE EARTH IS STILL NOT COMPLETED AFTER CENTURIES OF EXPLOR-ATION AND RESEARCH.

Will People Settle on Mars?

The end of the historic Mars 1 mission was not the end of the discovery of the red planet – many mysteries remain. But it was a major step. Bases were built on Mars itself and on its satellite Phobos, where the next explorers will be able to settle directly. When Michel Morey and his fellow astronauts return to Earth near the end of 2035, a new crew will already be in training. The departures of *Tsiolkovski 2* and *Tsiolkovski 3* are planned for June 2037. This time, 12 astronauts will leave for the red planet onboard two spacecraft.

New equipment is already headed to Mars onboard the *Oberth 3* and *Oberth 4* cargo vessels. The von Braun base is to be greatly expanded, and a second base will be set up in another area of the planet. Humanity's occupation of Mars has truly begun. However, it is only an intermittent occupation, like that of Earth's polar bases; the astronauts work on Mars a little over two years and then return to Earth, leaving the red planet uninhabited until the next expeditions. Will the red planet one day be occupied permanently by alternating teams in larger, more comfortable bases?

1

That stage is planned for 2050, when some astronauts will stay for over four years to act as a liaison between two successive missions. Four years on Mars! It will be a small step toward colonization of the red planet – a dream that so many space enthusiasts have held since the end of the 20th century. The experiment will be limited to six astronauts but with something completely new. The team will consist of three couples, despite the reservations of psychologists, who fear emotional complications, and those of physicians, worried about the possible birth of children. In principle, the births would not be authorized, but more than likely some astronauts will ignore the injunction in order to produce the first human beings born on another planet (the first "Martians"?).

Two physicians will be on the team for this first long-term stay on Mars, and it won't be any surprise if infant care equipment is sent to the red planet, just in case …

1 THE VON BRAUN BASE GROWS WITH EVERY EXPEDITION, WITH GIANT DOMES AND LONG, COVERED WALKWAYS CONNECTING SMALL MODULES WHERE THE FIRST COLONISTS WORK, RELAX AND LIVE.

2 SMALL SCIENTIFIC BASES ARE SET UP IN MORE INTERESTING AREAS OF THE PLANET …

3 …CONNECTING WITH EACH OTHER BY ALL-TERRAIN VEHICLES AND SHUTTLES.

Is Mars Habitable? What about Terraforming?

As soon as permanent settlers live on Mars after 2050, the arrival of babies will only be a question of time: One day, in the second half of the 21st century, a child and then children will be born on the red planet. The first ones will accompany their parents back to Earth, but humans may someday decide to stay on Mars despite the hardships of life under domes, in a world where the only excursions are made in spacesuits. For ages, human beings have demonstrated their ability to undertake voyages without return, to settle in faraway, often inhospitable lands. The conquest of Mars will be no exception. Before the end of the 21st century, dozens of men, women and children will be permanently settled on the red planet.

Can this cramped existence be transcended and Mars transformed into another Earth, with a breathable atmosphere, fields, forests, rivers, lakes and seas? In the 1950s, such a transformation was called "terraforming" by American writer Jack Williamson. It would be a gigantic undertaking, which could be called "planetary engineering." In effect, it involves nothing less than transforming a planet's environment from top to bottom.

In 1991, three American researchers, Chris McKay, Owen Toon and James Kasting, imagined the possible stages involved in terraforming Mars. First would be the creation of a denser, warmer atmosphere through the evaporation of the frozen carbon dioxide found in the polar ice caps. To initiate the process, the proposed approach is to produce enormous quantities of gases known as CFCs on the red planet and inject them into its atmosphere. These gases, prohibited from production on Earth, are very effective at absorbing infrared radiation and creating the greenhouse effect. After more than a century, they will give Mars an atmosphere 140 times denser than today and a mean temperature of 32°F (0°C). Its inhabitants could start going outside without spacesuits, using a simple oxygen mask, but would need to be warmly dressed.

Next would come the evaporation of frozen water at the poles and especially in the planet's subsoil, a stage that would take more than a century. Valles Marineris would gradually become an immense river flowing into an ocean occupying all the low plains surrounding the North Pole. The higher Southern Hemisphere would form a vast continent, with inland seas occupying large craters like Hellas and Argyre. The planet would no longer be red, but blue. Its atmosphere, by then as dense as Earth's, would still not be breathable, as it would lack oxygen, which only plants can release. So Mars would have to be made green by seeding it with genetically modified plants. This would again take centuries, but by the dawn of the fourth millennium, Mars might finally be another Earth, a second haven for humanity in the solar system!

This terraforming would demand a titanic effort, and its feasibility remains to be proven. Is it a good idea? Does humanity have the right to change a world for its own convenience? Shouldn't Mars remain the red planet with its originality and its own beauty? The solar system can offer humanity many other options for leaving Earth and settling in space.

1 MARS WITH A VAST NORTHERN OCEAN AND SEAS FOLLOWING ITS TERRAFORMING IN THE THIRD MILLENNIUM. 2 IF MARS BECAME "BLUE," THE GREAT EQUATORIAL PLAIN HELLAS, WHICH OCCUPIES THE BOTTOM OF A VAST CRATER ... 3 ... WOULD BE SUBMERGED BY WATER, AND LONG GULFS WOULD STRETCH TOWARDS HESPERIA PLANUM IN THE NORTH. 4 RIVERS WOULD FLOW INTO LAKES ... 5 ... AND THE IMMENSE CANYONS OF VALLES MARINERIS ... 6 ... WOULD BECOME GIGANTIC RIVERS.

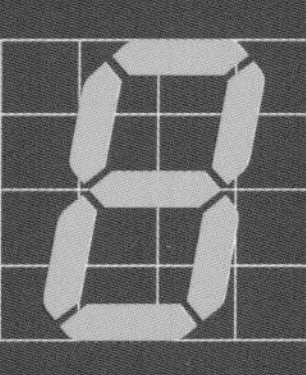

Will the Moon Be Colonized?

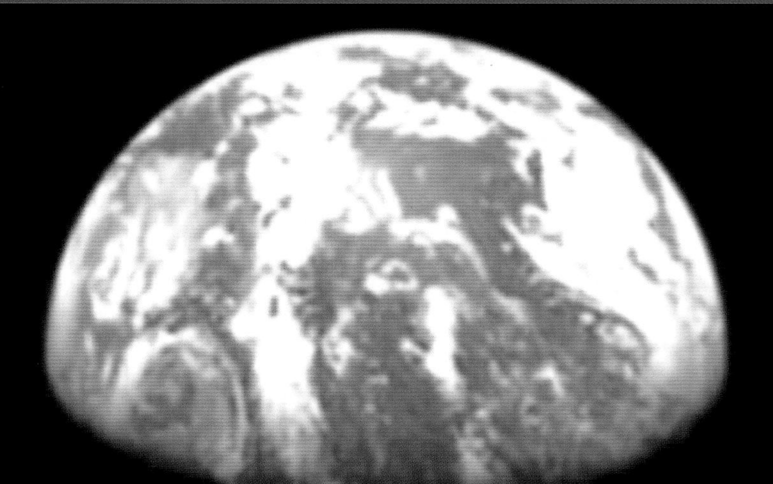

Unlike Mars, the Moon could never become another Earth. In the past, it lost any volatile substances (especially water and carbon dioxide) that could create an atmosphere. Only a little ice, brought by comets over the ages, doubtlessly remains at the bottom of craters close to the poles, where perpetual darkness reigns, cut off from any sunshine. This ice was detected for the first time in 1994 by the small American probe *Clementine*. It is a considerable resource on the Moon's surface but could never be used as a basis for terraforming. The Moon will remain forever a mineral desert, plunged into the cosmic void. But will that prevent astronauts from occupying it permanently or even colonizing it?

Astronauts returned to the Moon in the 2020s to conduct scientific research and prepare for the *Tsiolkovski*'s Martian expedition. However, that only involved setting up a small forward base that uses telescopes and radio telescopes installed on the dark side, protected from Earth-based interference. Might it be possible to go a step further and excavate in the lunar soil, sheltered from harmful solar particles and cosmic radiation, habitats where people could live comfortably and even settle permanently? It's not impossible but highly unlikely. The lunar environment is especially hostile, distinguished by two week-long freezing nights alternating with blazingly hot days that are just as long, where light is so intense and shadows so dark it is very hard to make out any kind of terrain and move around. Further, the lunar landscapes lack grandeur and beauty.

Living on the Moon would be tantamount to a caveman's existence, with rare excursions in pressurized vehicles. Moreover, Earth is not far away – just a few days' travel. So why would people ever want to settle on the Moon and raise families? For the pleasant experience of low gravity, six times less than Earth's, making it easy to move around and practice new sports? Or because of an attraction to a world that might become, after 2050, especially in 2100, an industrial base for extracting from the soil materials used to later build the spacecraft, rockets and habitats needed to expand human existence into space? It would actually be easier and cheaper to conquer space using lunar materials rather than those from Earth because of the low gravity on the Moon. But other objects, such as small asteroids, would offer even cheaper sources of material. How many humans will be living on the Moon in 2100? No doubt a few hundred will stay there for work shifts of a few months, a little like people who work on oil drilling platforms at sea, but not many others.

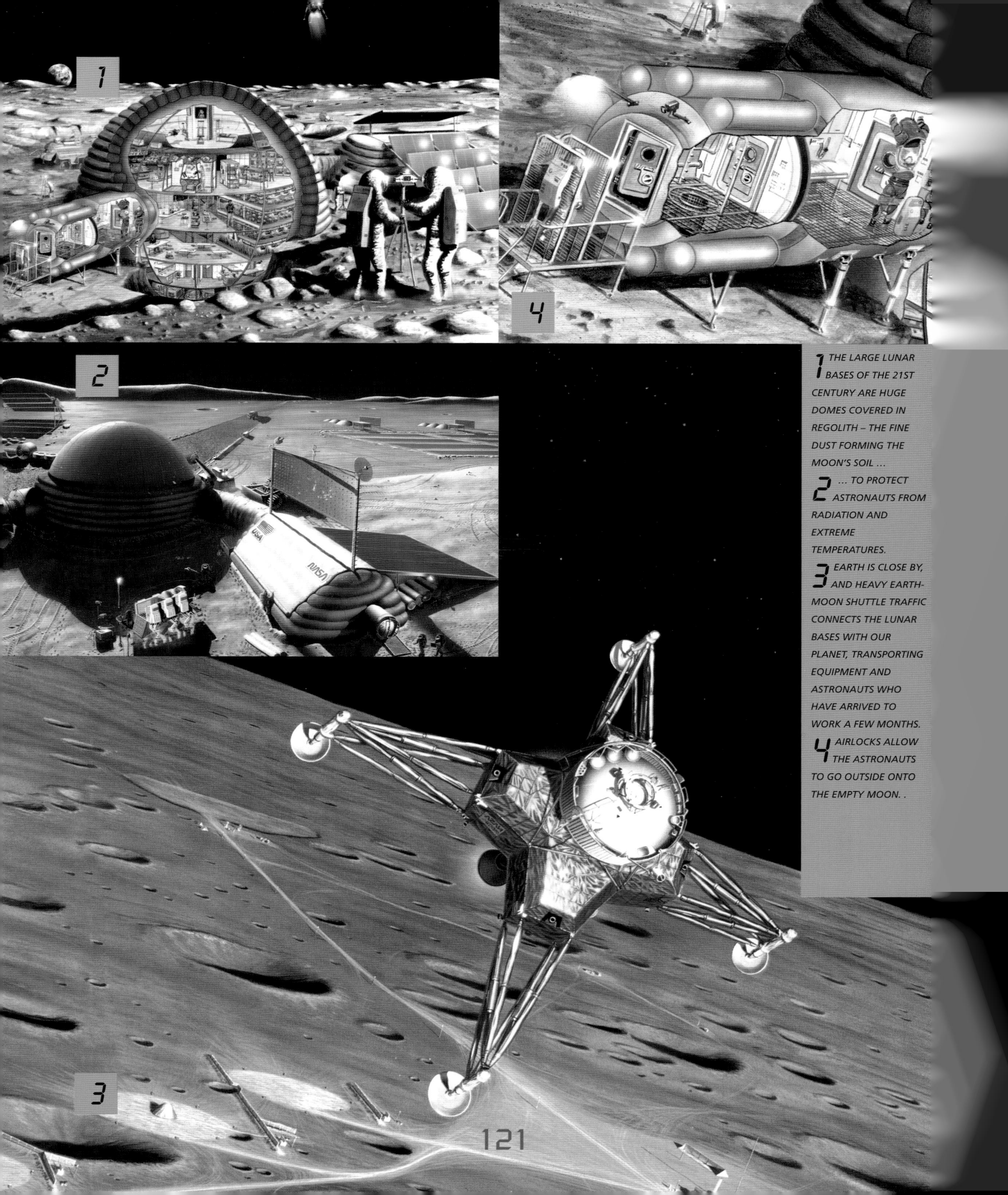

1 THE LARGE LUNAR BASES OF THE 21ST CENTURY ARE HUGE DOMES COVERED IN REGOLITH – THE FINE DUST FORMING THE MOON'S SOIL ...

2 ... TO PROTECT ASTRONAUTS FROM RADIATION AND EXTREME TEMPERATURES.

3 EARTH IS CLOSE BY, AND HEAVY EARTH-MOON SHUTTLE TRAFFIC CONNECTS THE LUNAR BASES WITH OUR PLANET, TRANSPORTING EQUIPMENT AND ASTRONAUTS WHO HAVE ARRIVED TO WORK A FEW MONTHS.

4 AIRLOCKS ALLOW THE ASTRONAUTS TO GO OUTSIDE ONTO THE EMPTY MOON. .

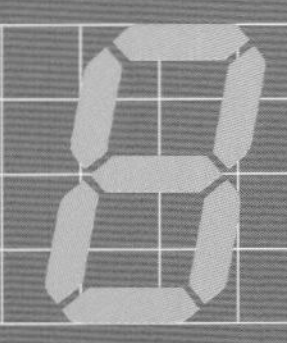

Space Habitats

Mars is a long way off and, at best, will only be habitable in several centuries. The Moon is close by but very inhospitable. Where, then, could colonists anxious to live in space settle close enough to Earth and in comfortable environments? The American physicist, Gerard O'Neill came up with an original solution in the mid-1970s: Build habitats right in space, away from any planetary surface! They would be like gigantic orbiting stations where living conditions reminiscent of those on Earth, and in some ways more pleasant, could be created onboard.

The Earth-Moon system has locations with remarkable properties. They are the five Lagrange Points, named after the famous French mathematician of the 18th century, where an object is permanently fixed in the same spot in relation to Earth and the Moon. One of these points, L2, is 40,000 miles (65,000 km) above the dark side of the Moon, in an ideal position for a base that could receive lunar materials intended for industrial use. Two others, L4 and L5, are on the same trajectory as the Moon, an average of 240,000 miles (385,000 km) from Earth, but respectively five days ahead and behind the Moon's movement.

Gerard O'Neill felt that L4 and L5 were perfect locations for establishing colonies. The initial idea was to build gigantic cylinders using materials extracted from the Moon. Space colonists would live on the inside surface of the cylinders, which would spin around to recreate artificial gravity using centrifugal force. Every colony would have large windows that let in sunshine at a normal diurnal rhythm, along with artificial landscapes of forests, fields and rivers, with small buildings constructed in the middle.

The smallest colony would be 325 feet (100 m) in diameter, over half a mile (1 km) long and have 100,000 inhabitants! The largest would be 10 times that size and could become a permanent home for 10 million men, women and children. The Sun would provide the necessary energy. Hydroponic (soil-free) crops and livestock would provide food. The climate would be perfectly controlled and always pleasant.

An idyllic life, far from earthly worries – why not? But who would finance the construction of these enormous infrastructures, weighing hundreds of thousands of tons? The space colonies would be industrial centers making use of lunar materials to build equipment to serve Earth societies: rockets, satellites and spacecraft, but also "solar power stations" in space, providing Earth with its own abundant energy. Is it the impossible dream of a visionary or the royal road to the colonization of space? The third millennium will answer that question, but colonies like those imagined by Gerard O'Neill could one day, far in the future, harbor even more people than Earth.

1

1 LIKE A GIGANTIC WHEEL SPINNING IN THE IMMENSITY OF SPACE, A SPACE COLONY IN THE FORM OF A TORUS, WITH THOUSANDS OF INHABITANTS LIVING IN ARTIFICIAL GRAVITY, BASKS IN SUNSHINE WHILE FAR OUT IN SPACE.

2 ENORMOUS CYLINDRICAL COLONIES IMAGINED BY GERARD O'NEILL ARE LINKED TWO-BY-TWO, WITH CROP GROWING AREAS RESEMBLING STRINGS OF PEARLS.

3 LANDSCAPES RESEMBLING THOSE ON EARTH ARE RECREATED INSIDE TORUS-SHAPED SPACE COLONIES ...

4 ... LIKE THOSE THAT ARE EVEN LARGER, IN THE FORM OF CYLINDERS ...

5 ... WHERE GRAND VISTAS OPEN UP FOR MILLIONS OF INHABITANTS.

8 The Industrialization of Space

Permanent human settlement in space, abundant solar energy, materials available from the Moon or asteroids ... will space in the 22nd century be the site of a new industrial revolution? Such a revolution, bestowing economic importance on space, would doubtlessly accelerate the expansion of humanity beyond Earth. But will that really happen? Besides power and materials, other elements might hasten its arrival, starting with the absence of pollution. Unlike our planet, space is infinite, and both heat and waste could be disposed of without harming the Earth's environment. The surrounding emptiness, conducive to many industrial processes, and weightlessness are also favorable factors.

However, the value of certain constituents of extraterrestrial materials will perhaps be the determining factor. Such is the case, for example, of helium-3 atoms, a very rare form of helium found in the upper layer of lunar soil. These atoms come from the Sun, whose particles directly bombard the lunar surface. They have been accumulating for millions of years at a rate of some millionths of grams per mile of soil.

It's not much, but even in minuscule quantities helium-3 is a choice product. In a few decades it could become the preferred fuel for thermonuclear reactors, using the same principle as hydrogen bombs or the source of the Sun's power – thermonuclear fusion. At the energy prices in the year 2000, 1 pound (0.45 kg) of helium-3 would be worth $6.8 billion. The reserves of lunar helium-3 could satisfy humanity's power needs for millennia! There is one problem, however. To extract it, the entire Moon would have to be turned into a gigantic open pit mine. Will people really want to disfigure this celestial body that they have admired in the heavens since time immemorial?

Some asteroids, average-sized rocks traveling between major planets, would also be enormously valuable. These are the ferrous asteroids that represent approximately 3 percent of asteroids, and are almost entirely formed of ferrous metals such as iron and nickel, but also gold and platinum. A ferrous asteroid over half a mile (1 km) in diameter could supply Earth with iron for over 10 years, and nickel for a millennium. It would also contain some 100,000 tons of platinum and 10,000 tons of gold. At market prices in the year 2000, it would be worth $1 trillion. The problem would obviously be moving it close to Earth and then onto the planet.

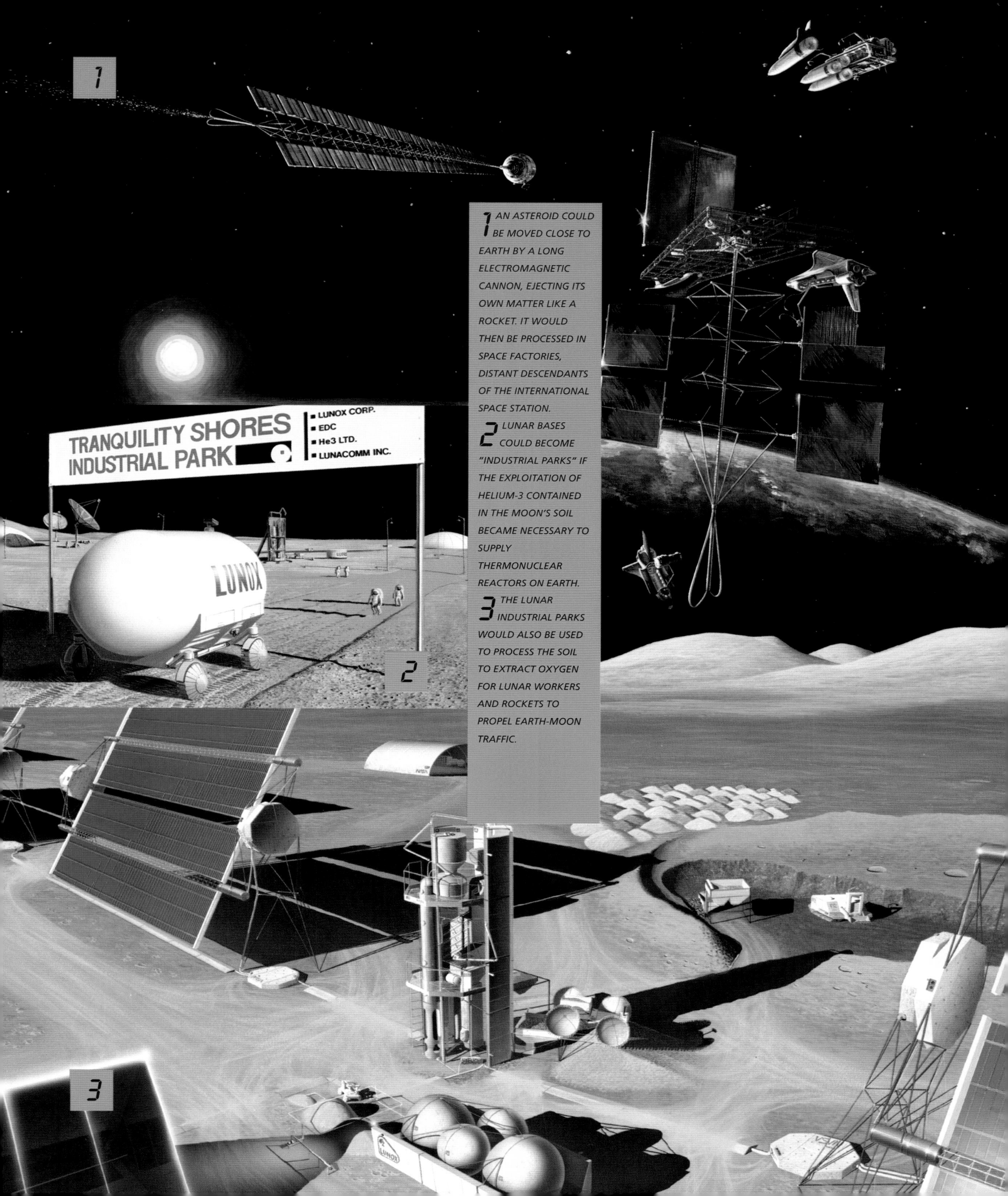

1 AN ASTEROID COULD BE MOVED CLOSE TO EARTH BY A LONG ELECTROMAGNETIC CANNON, EJECTING ITS OWN MATTER LIKE A ROCKET. IT WOULD THEN BE PROCESSED IN SPACE FACTORIES, DISTANT DESCENDANTS OF THE INTERNATIONAL SPACE STATION.

2 LUNAR BASES COULD BECOME "INDUSTRIAL PARKS" IF THE EXPLOITATION OF HELIUM-3 CONTAINED IN THE MOON'S SOIL BECAME NECESSARY TO SUPPLY THERMONUCLEAR REACTORS ON EARTH.

3 THE LUNAR INDUSTRIAL PARKS WOULD ALSO BE USED TO PROCESS THE SOIL TO EXTRACT OXYGEN FOR LUNAR WORKERS AND ROCKETS TO PROPEL EARTH-MOON TRAFFIC.

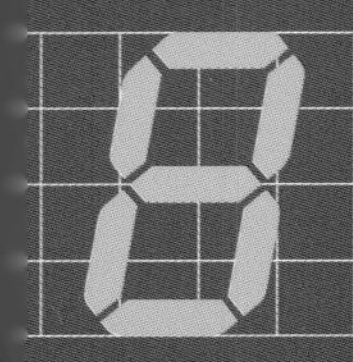

Humanity's New Realm

With space flights around Earth and the Apollo missions in the second half of the 20th century, humanity expanded its sphere of activity to Earth's outer reaches of space, to that part of space dominated by the attraction of its home planet. With the voyage of the *Tsiolkovski*, human civilization became planetary and started expanding into what will become its sphere of influence throughout the third millennium: the solar system.

The solar system is primarily the domain of the planets – nine bodies worthy of this name, with their procession of satellites (some one hundred in all) – the Moon, Mercury, Venus, Earth, Jupiter, Saturn, Uranus, Neptune and Pluto. In 2035, humanity will have already established a foothold on two of these planets. But that's only the beginning. For decades and then centuries, astronauts will discover all the planets and set up bases or even colonies on their soil, when it exists, or on that of their satellites. A society and an interplanetary economy will gradually establish itself. The globalization phenomenon that marked the 20th century will thus spread to planetary worlds.

But the solar system does not stop at Pluto or Neptune – the planets that are farthest from the Sun, depending on their periods, at a distance some 35 times farther than Earth's. The solar system stretches much farther into interstellar space, with the icy asteroids of the Kuiper Belt and beyond to the comets in the Oort Cloud, which may reach up to one light-year in distance. This huge realm is now within the grasp of humanity and its ambitions.

1 NINE PLANETS COMPOSE THE NEW SPHERE OF HUMANITY'S ACTIVITY IN THE THIRD MILLENNIUM.

2 LIKE CHRISTOPHER COLUMBUS AND HIS THREE SHIPS SETTING OUT TO DISCOVER AMERICA, HUMANS WILL FOLLOW INTERPLANETARY ITINERARIES.

3 CHILDREN OF THE 21ST CENTURY ALREADY FEEL AT HOME IN INTERSTELLAR SPACE, WHERE THEY LEAVE ON VIRTUAL TRIPS BEFORE ACTUALLY DEPARTING FOR THE OTHER PLANETS.

2

3

1

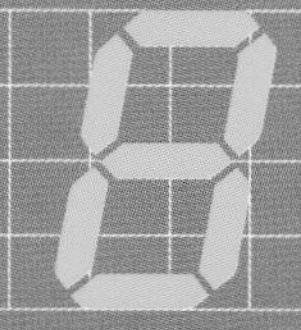

The Sun

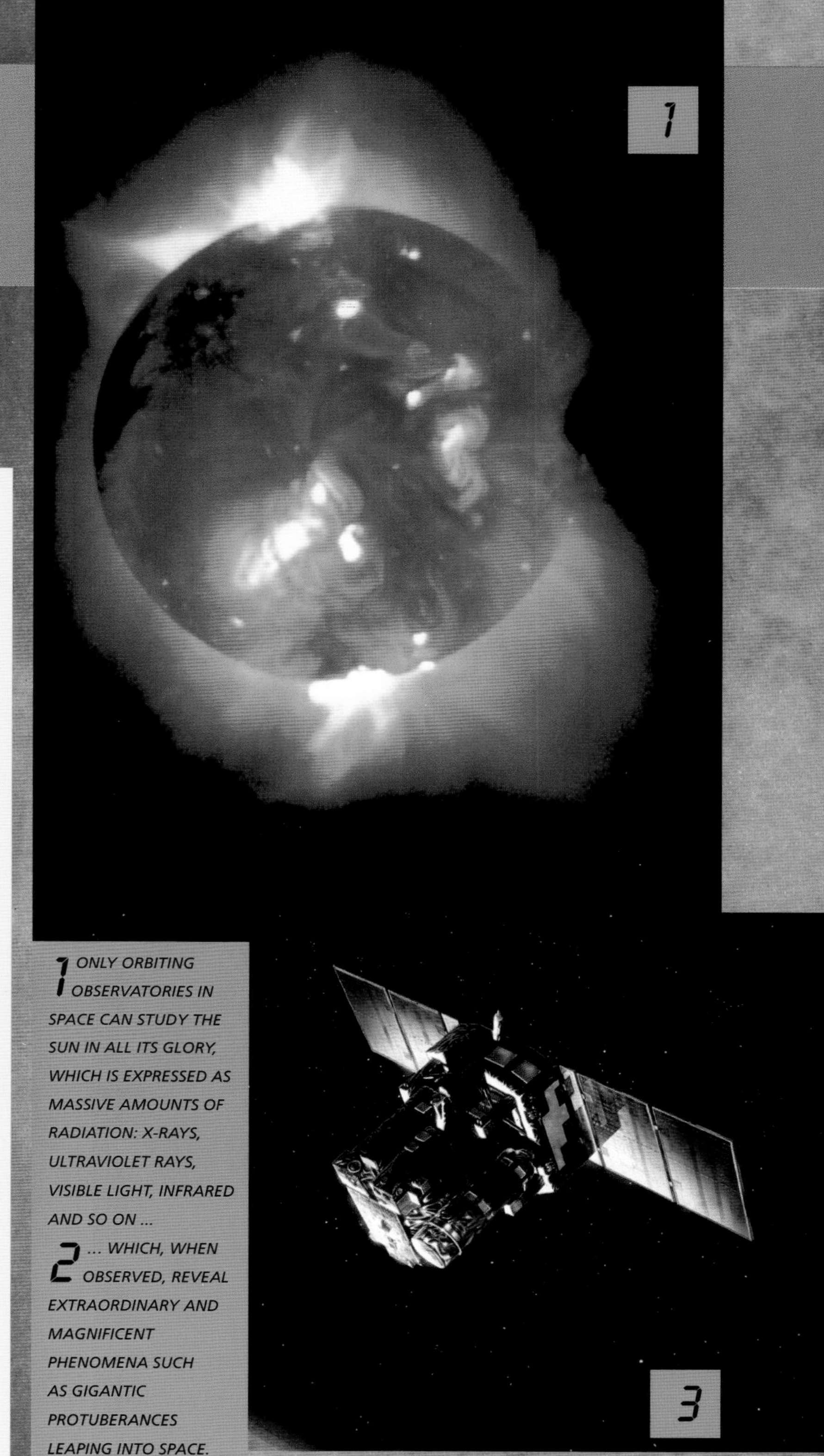

Without the Sun there would be no solar system. Since its formation some 4.5 billion years ago, the Sun has guided the formation of the planets and served as the driving force of their evolution. Its radiation and heat drove out the volatile atoms and molecules that today form the basic elements of the "giant planets" of Jupiter, Saturn, Uranus and Neptune. The same phenomenon left rocky bodies formed mainly of silicates, in areas close to the Sun, gaving rise to the "telluric" planets of Mercury, Venus, Earth and Mars. The Sun's heat provided the energy that enabled life to evolve on Earth – and perhaps on Mars. Without it, there would be no life. But, on the other hand, too much heat excludes all life, as is the case on Venus. Our planet's climate is optimal for terrestrial life, so we are responsible for controlling our industrial activities if we are to prevent the "greenhouse effect," which causes retention of the Sun's heat.

Seen from afar, perhaps by extraterrestrials living dozens or even hundreds of light-years away, the Sun is a very ordinary star because of its mass and type, located halfway along the "Main Sequence" of stellar evolution. The Sun will continue to burn hydrogen in its interior and thus form helium for more than four billion years. After that, it will start to consume heavier and heavier nuclear matter in its own "thermonuclear furnace" until it has exhausted all of its enormous resources. That will spell the end of the Sun – and the Earth, which will be absorbed and reduced to ash by a Sun that has become a red giant. Will humanity be concerned about this spectacular end? Probably not: If the human race has not disappeared, its descendants will have left Earth a long time before the cataclysmic event.

1 ONLY ORBITING OBSERVATORIES IN SPACE CAN STUDY THE SUN IN ALL ITS GLORY, WHICH IS EXPRESSED AS MASSIVE AMOUNTS OF RADIATION: X-RAYS, ULTRAVIOLET RAYS, VISIBLE LIGHT, INFRARED AND SO ON ...

2 ... WHICH, WHEN OBSERVED, REVEAL EXTRAORDINARY AND MAGNIFICENT PHENOMENA SUCH AS GIGANTIC PROTUBERANCES LEAPING INTO SPACE.

3 THE EUROPEAN SOHO OBSERVATORY HAS GREATLY INCREASED OUR KNOWLEDGE OF THE SUN.

8 Mercury

1

2

Mercury is remarkable for its proximity to the Sun. This also makes it hard to observe, as it's only visible for a short time after sunset or just before dawn, and then only at certain times of the year. It is not very large – only Pluto is smaller. Mercury's mass is only one-twentieth that of Earth's, and it's completely devoid of an atmosphere. Mercury looks surprisingly like the Moon – a globe pockmarked with craters. In 1974, it was examined for the first time in detail by the *Mariner 10* space probe. Temperature fluctuations there are very extreme: 770°F (450°C) on the side illuminated by the Sun and –275 (–170°C) in the shade. These are the largest extremes in the solar system. This world is even more inhospitable than the Moon. Space exploration of Mercury ended with *Mariner 10* in the 20th century. Why continue studying a planet that did not seem to offer anything very exciting? Will things be different in the 21st century? Accurate data on all the planets are needed to understand how the solar system was formed and how it's evolving. In 2009, the ESA will launch a Mercury probe called *Bepi Colombo*, with one orbiter and two landers, and more ambitious projects are being studied to bring back soil samples. But will humanity set foot on this planet? It doesn't really seem necessary even though some Earth-based observations appear to reveal that on Mercury, as on the Moon, ice could exist in polar craters sheltered from the Sun. That means a base could be established in such areas one day and be used, for example, as an observatory to study the Sun up close.

1 MERCURY AS SEEN BY MARINER 10: A WORLD SCARRED BY ASTEROID IMPACTS DURING THE FIRST BILLION YEARS OF THE SOLAR SYSTEM.

2 LAVA-COVERED PLAINS PRESENT A MUCH LESS FRACTURED SURFACE.

3 FIRST PICTURE OF MERCURY TAKEN BY MARINER 10 ON MARCH 24, 1974, FROM 3,340,000 MILES (5,380,000 KM) AWAY.

4 THIS CHAOTIC AREA WAS FORMED BY A GIGANTIC IMPACT THAT CREATED MERCURY'S LARGEST CRATER ON ITS UNDERSIDE, THE CAROLIS BASIN.

Venus

Until the 20th century, astronomers believed that Venus was Earth's sister planet with a tropical climate hidden under a permanent cloud cover – hotter than our planet's because of its proximity to the Sun. True, Venus is just slightly smaller than Earth and only 30 percent closer to the Sun, but space exploration has revealed a world completely different from Earth, where no life could exist. How can two celestial bodies so close to each other – the blue planet inhabited by people and the hellish Venus – be so different?. Conditions on Venus, in fact, are extreme: a temperature of 770°F (450°C) and pressure a hundred times greater than on the Earth's surface. Moreover, the clouds obscuring the Sun are composed of droplets of sulfuric acid. A few spacecraft landed on Venus following the Soviet probe *Venera 7*, the first to do so in 1970. Some sent back photos from the surface showing volcanic rocks bathed in a faint reddish light, but none of them survived for more than two hours in the planet's extreme environment. Other craft outfitted with imaging radar, in particular NASA's *Magellan* probe, scanned the surface of Venus through the clouds and made three-dimensional maps.

Venus has revealed itself to be a volcanic planet where 80 percent of the soil is covered in lava. The next step, in the 21st century, will be to bring back soil samples. Such an operation, planned by the European Space Agency, will be difficult: Venus' attraction is almost as strong as Earth's, and escaping from it in a dense, superheated atmosphere will truly be challenging.

Will people ever land on the surface of this burning planet? Most likely not. At best, manned spacecraft will fly past Venus on the way to other regions of the solar system. There are Russian projects that speak of piloted balloons able to float in the upper atmosphere, but why? When it comes to ideas about "terraforming" Venus, nothing very serious has been advanced.

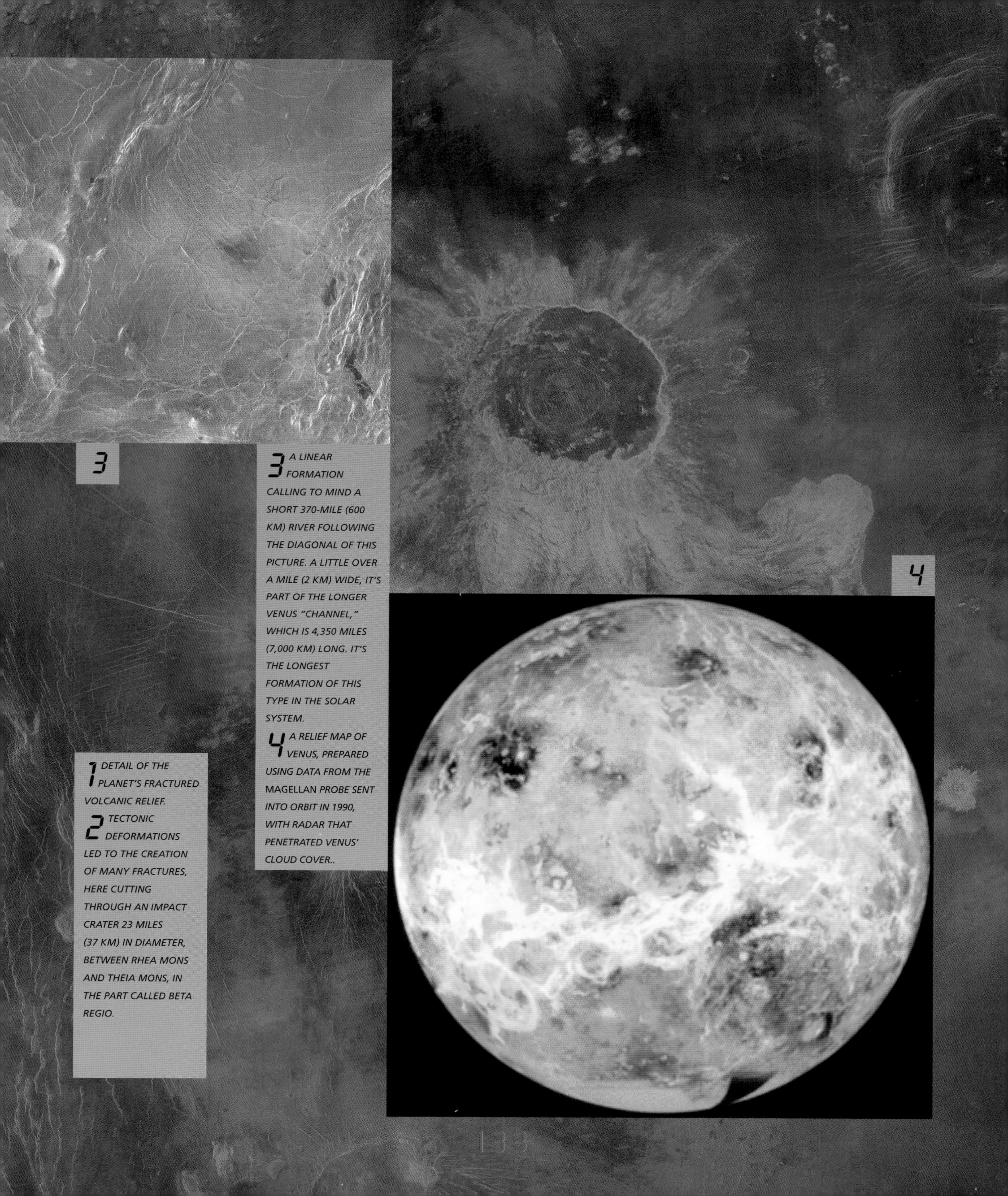

1 DETAIL OF THE PLANET'S FRACTURED VOLCANIC RELIEF.

2 TECTONIC DEFORMATIONS LED TO THE CREATION OF MANY FRACTURES, HERE CUTTING THROUGH AN IMPACT CRATER 23 MILES (37 KM) IN DIAMETER, BETWEEN RHEA MONS AND THEIA MONS, IN THE PART CALLED BETA REGIO.

3 A LINEAR FORMATION CALLING TO MIND A SHORT 370-MILE (600 KM) RIVER FOLLOWING THE DIAGONAL OF THIS PICTURE. A LITTLE OVER A MILE (2 KM) WIDE, IT'S PART OF THE LONGER VENUS "CHANNEL," WHICH IS 4,350 MILES (7,000 KM) LONG. IT'S THE LONGEST FORMATION OF THIS TYPE IN THE SOLAR SYSTEM.

4 A RELIEF MAP OF VENUS, PREPARED USING DATA FROM THE MAGELLAN PROBE SENT INTO ORBIT IN 1990, WITH RADAR THAT PENETRATED VENUS' CLOUD COVER..

The World of Jupiter

The Ancients were right in naming this giant after the master of Olympus. Jupiter is by far the king of the planets, the only one extraterrestrials could make out around the Sun from dozens or hundreds of light-years away. Jupiter alone contains two-thirds of the total mass of the planets. But is it really a planet? Could it just be a failed star instead? True, its composition is close to that of the Sun's – mainly hydrogen and helium. Had it been 10 times more massive, thermonuclear reactions would have been triggered in its interior and transformed it into a star. The Sun would then have been a double star, and the fate of the solar system, and that of the Earth, would have been completely different. But that didn't happen, and Jupiter is just a very large planet.

Trailing along with Jupiter, as it circles the Sun, is a procession of no less than 63 satellites. Four of them helped change the history of science and civilization. In 1610, Galileo pointed his first telescope toward Jupiter and observed a small star-like system around it, composed of Io, Europa, Ganymede and Callisto.

1 JUPITER AND ITS FOUR MAIN SATELLITES: CALLISTO, IN THE FOREGROUND, FOLLOWED BY GANYMEDE, EUROPA AND, FINALLY, IO.

2 THE AMERICAN SPACECRAFT GALILEO, PHOTOGRAPHED IN 1989 FROM THE SPACE SHUTTLE ATLANTIS JUST AFTER ENTERING EARTH'S ORBIT AND BEFORE HEADING OUT ON A LONG SIX-YEAR TRIP TO JUPITER.

3 SECOND HALF OF THE 21ST CENTURY, ASTRONAUTS EXPLORING EUROPA, BATHED IN JUPITER'S LIGHT.

More than any other, this discovery proved that Ptolemy's model of the universe was wrong and that the planets are celestial bodies like the Earth, circling the Sun, and could have their own satellites.

People will never set foot on Jupiter's soil simply because Jupiter has no surface! Like the other giant planets (Saturn, Uranus and Neptune), it's a ball of gas whose atmosphere gets hotter and denser the deeper it is penetrated. Hydrogen gradually becomes a liquid and then a solid, and a core of silicates and metals most certainly occupies the center. The only thing visitors to Jupiter could do would be to float in balloons in the upper atmosphere, and only if they had survived the trip through the belts of radiation circling the planet, which could kill a person in less than an hour. Astronauts will certainly travel toward Jupiter, but their actual goal will be one of its satellites, Europa, one of the most mysterious and promising objects in the solar system.

8 Europa's Ocean

What if Mars were not the best place to look for life in the solar system? Europa, smaller than the red planet and much farther from the Sun, might provide very favorable conditions. The second farthest satellite from Jupiter, Europa is no larger than the Moon, and the Sun's rays, 25 times less intense than on Earth and 12 times less than on Mars, would not make its surface habitable. Any potential life would emerge not on, but under the surface, which would be nothing more than a giant ice pack covering an immense ocean at a thickness of a few dozen miles. For a long time, astronomers wondered whether Europa was an ice-covered world, like Ganymede (the largest of Jupiter's satellites) and Callisto.

So it was no major leap to go from that concept to imagining that a liquid ocean exists below the surface, especially after seeing the fantastic images transmitted by NASA's *Galileo* probe orbiting Jupiter since 1995. These images reveal a young, practically crater-free surface, as if water coming from the ocean regularly made use of cracks and geysers to renew the ice pack. These images also show overlapping slabs of ice, fractures and even structures resembling icebergs.

The thickness of the ice pack, a few miles or perhaps a few dozen miles on average, would only be several dozen feet in some locations.

The energy needed to melt the ice of Europa and sustain this immense subsurface ocean would be freed up by movements generated in the core and the ice by gigantic tides caused by the attractions of Jupiter, Ganymede and Callisto. Sub-marine volcanoes might even expel chemical elements into Europa's ocean, made up of nutrients that could give rise to and sustain a form of life. Arthur C. Clarke imagined such a life form in his book *2010: Odyssey Two*, a sequel to the celebrated *2001: A Space Odyssey*. But how can anyone search for life in an ocean buried under miles of ice, hundreds of millions of miles from Earth? A first step will be placing a satellite in orbit around Europa. Robots would then have to be sent down that could penetrate the ice and dive into the ocean with lights, cameras and a whole series of measuring devices before astronauts would have their turn at exploring Europa, only one or two decades after stepping off the *Tsiolkovski* and onto the red planet.

1

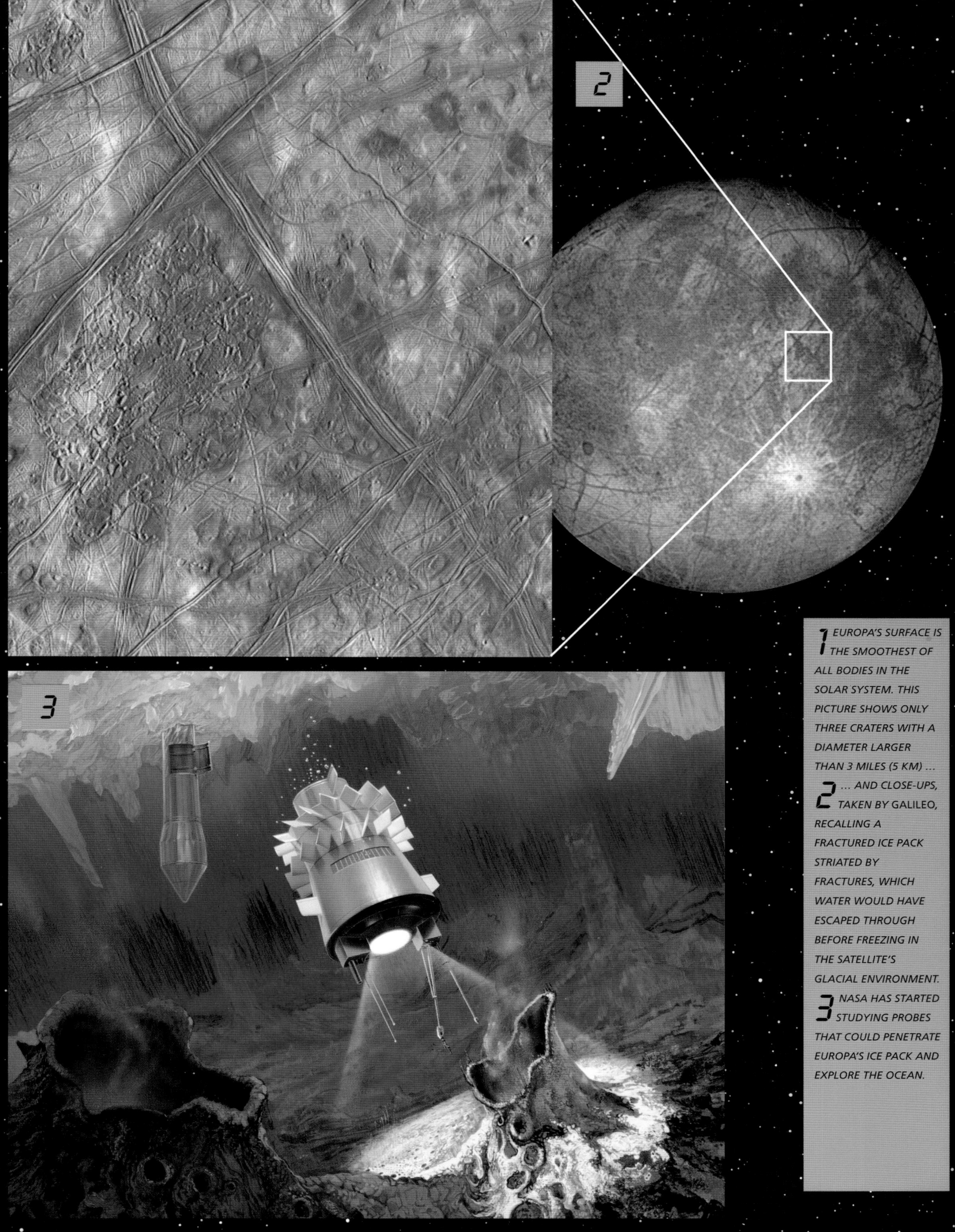

1 EUROPA'S SURFACE IS THE SMOOTHEST OF ALL BODIES IN THE SOLAR SYSTEM. THIS PICTURE SHOWS ONLY THREE CRATERS WITH A DIAMETER LARGER THAN 3 MILES (5 KM) …

2 … AND CLOSE-UPS, TAKEN BY GALILEO, RECALLING A FRACTURED ICE PACK STRIATED BY FRACTURES, WHICH WATER WOULD HAVE ESCAPED THROUGH BEFORE FREEZING IN THE SATELLITE'S GLACIAL ENVIRONMENT.

3 NASA HAS STARTED STUDYING PROBES THAT COULD PENETRATE EUROPA'S ICE PACK AND EXPLORE THE OCEAN.

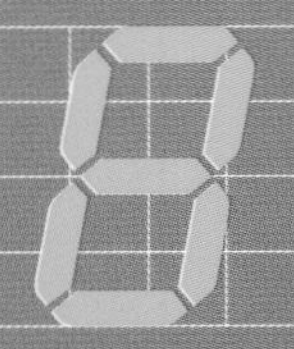

Io's Volcanoes

Exploration often leads to completely unexpected discoveries. The space exploration of Io, the closest Galilean satellite to Jupiter, is a magnificent example. Like Europa, Ganymede and Callisto, astronomers considered Io a satellite eternally covered in ice, which would be in no danger of being melted by the Sun's weak rays. But photos taken in 1979 by NASA probes *Voyager 1* and *Voyager 2* completely overturned this conception. Io is not covered in ice, but sulfur expelled by very active volcanoes. About 10 volcanic vents were observed by the *Voyager* probes and then *Galileo*, propelling sulfur jets hundreds of miles high. Every year, Io's volcanoes produce 10 billion tons of sulfuric products that settle on the surface to create a layer 0.04 inches (1 mm) thick. The result is that Io has the youngest surface of any body in the solar system.

To their great surprise, astronomers therefore discovered that Io was, aside from Earth, the only volcanically active body in the solar system (though there are still doubts about Mars and Venus, which at any rate would experience much less intense activity). This phenomenon arises from the same source as that which would melt Europa's ocean – tides generated by the attraction of neighboring bodies, in this case Jupiter and Europa. The sulfur and sulfur oxides liquefied by the heating caused by tides escape as geysers. The very bright, stunning colors of the sulfur products give Io its extraordinary palette of yellow, red, orange and black. Io has a strange beauty, but no human being will be able to visit its surface to admire the hills and geysers of sulfur, as the satellite is located in Jupiter's extremely dangerous radiation belts.

1 MOSAIC OF IMAGES OF IO TAKEN BY THE AMERICAN VOYAGER PROBE, REVEALING AN ORANGE-COLORED SURFACE RIDDLED WITH VOLCANIC VENTS.

2 CHANGES ON IO'S SURFACE BETWEEN OCTOBER 1999 AND FEBRUARY 2000 CAUSED BY SULFURIC LAVA FLOWS.

3 TOHIL MONS, AN IONIAN VOLCANO..

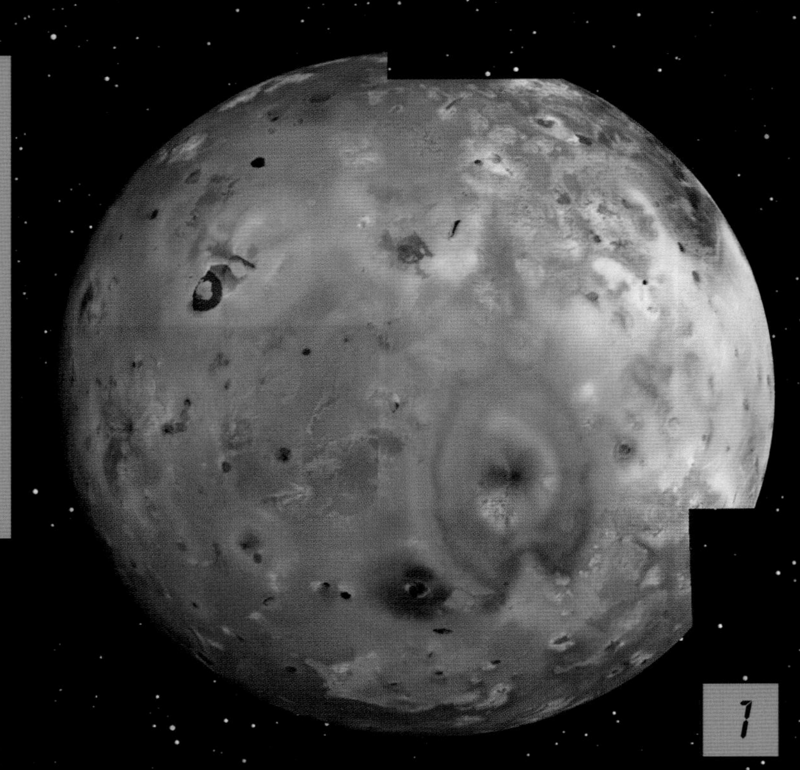

1

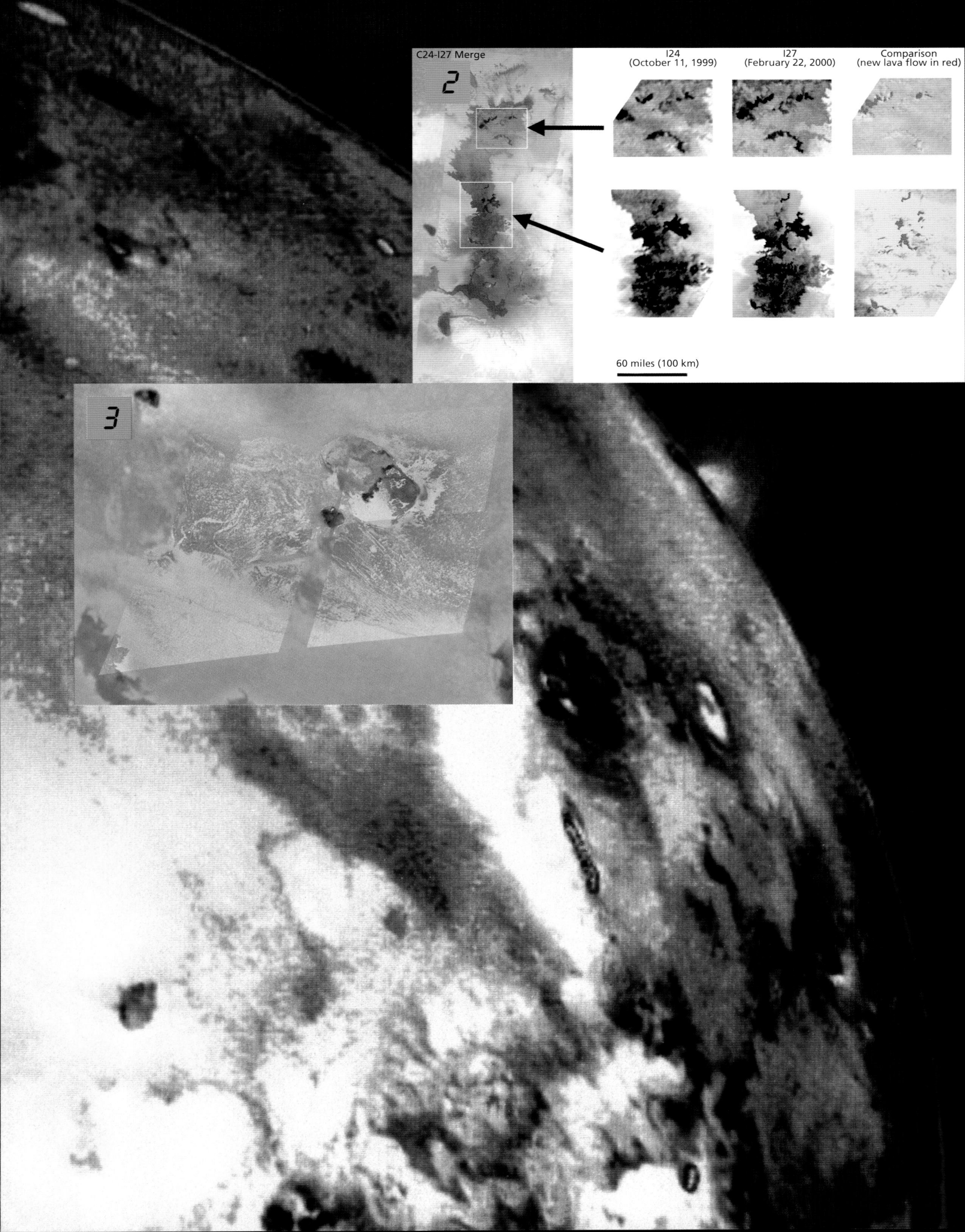

C24-I27 Merge
2
I24
(October 11, 1999)
I27
(February 22, 2000)
Comparison
(new lava flow in red)
60 miles (100 km)
3

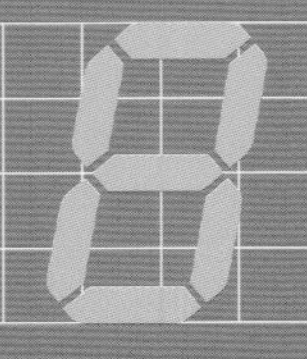

Lord of the Rings

Astronauts who reach Saturn – 10 times farther from the Sun than Earth – in the second half of the 21st century will be unable to land on the planet because, like Jupiter, Uranus and Neptune, it has no surface. Saturn is a ball of gas composed mainly of hydrogen, with a rocky core half as dense as Jupiter's. If a gigantic sea existed on Jupiter (the planet's diameter is 75,000 miles/120,000 km) Saturn could float in it! However, travelers approaching Saturn will not be disappointed. With its dazzling rings, the planet will give them the most fantastic views in the solar system. Discovered in the 17th century by the Dutch astronomer Christian Huygens, their awe-inspiring beauty was not truly revealed until the arrival of images taken by NASA's *Voyager* probes, which flew by Saturn in 1980 and 1981. The main rings are formed of millions of chunks of ice, broken up into thousands of smaller rings. They're very wide: 168,000 miles (270,000 km) in all, but no more than a few miles thick.

7

As is the case with Jupiter, the first visitors from Earth will head for the satellites of Saturn. Saturn has 30 satellites, seven of which are over 600 miles (1,000 km) in diameter. Some of them are very strange, like Iapetus, which has a dark side and bright side that inspired Arthur C. Clarke's *2001: A Space Odyssey* (in the film, Iapetus is the doorway to other universes established by extraterrestrials). Reality is more mundane, but Saturn's satellites certainly still hold many surprises for their explorers. The largest of them, Titan, is one of the most mysterious objects in the solar system.

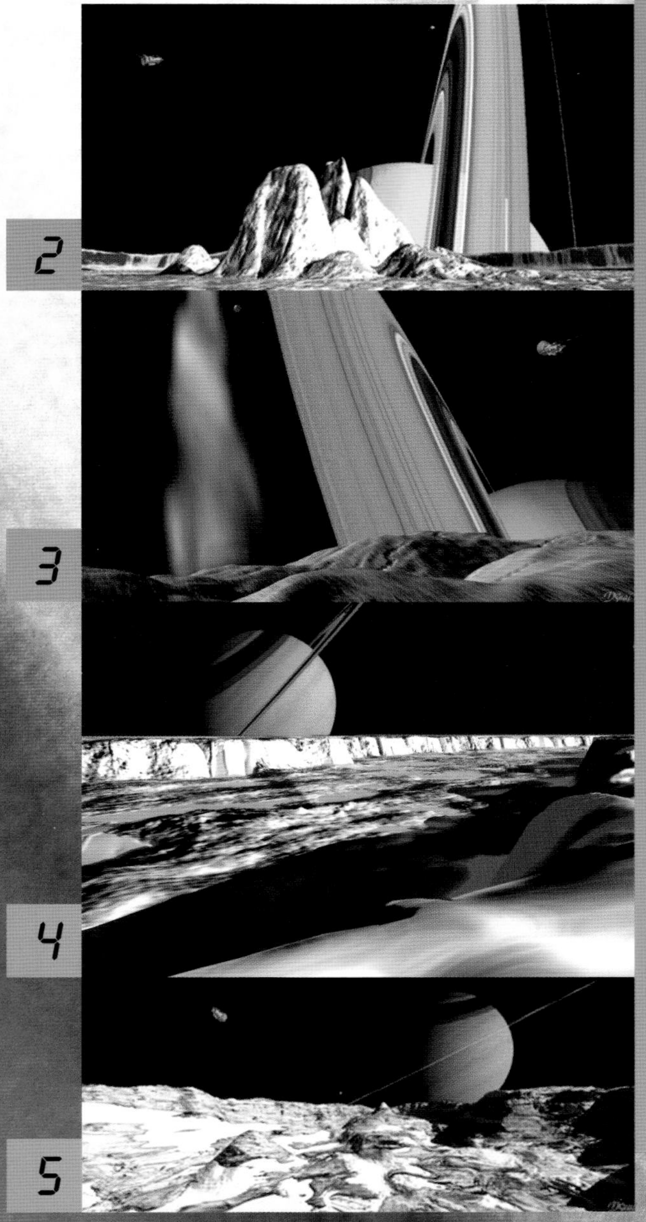

1 MONTAGE OF PICTURES OF SATURN AND SIX OF ITS SATELLITES: IN THE FOREGROUND, DIONE, WITH RHEA AND ENCELADUS ON THE RIGHT, TETHYS AND MIMAS IN THE BACKGROUND ON THE LEFT, AND, FINALLY, TITAN IN THE FAR UPPER RIGHT.

2 IMAGINARY LANDSCAPES ON SATURN'S SATELLITES – ON MIMAS IN THE IMMENSE HERSCHEL CRATER ...

3 ... AND THEN ON PANDORA, A SMALL, VERY IRREGULAR SATELLITE IN THE "F-RING," DISCOVERED BY THE AMERICAN VOYAGER PROBES, ...

4 ...AND ON TETHYS IN THE LONG ITHACA CHASMA TRENCH, EXTENDING FROM ONE POLE TO THE OTHER ...

5 ...WHERE THE SPLENDOR OF SATURN'S RINGS STILL DOMINATES THE SKY, SEEN HERE FROM THE CHAOTIC ICY SURFACE OF DIONE ...

6 ...AND HERE, FROM AN ICE GEYSER ON ENCELADUS.

7 AN ICE CLIFF ON HYPERION.

8 ON THE DARK SIDE OF IAPETUS WITH SATURN AND THE SUN IN THE SKY, 10 TIMES SMALLER THAN SEEN FROM EARTH.

9 AT THE BOTTOM OF A CREVASSE ON PHOEBE.

8 The Mysteries of Titan

Titan is remarkable for its size. Larger than Mercury, it's the second largest satellite in the solar system in terms of diameter, just after Ganymede. But what makes Titan really extraordinary is its atmosphere, which will probably make it the third celestial body to be explored by astronauts, after Mars and Europa. This atmosphere is denser than Earth's; surface pressure there is 30 percent higher than that on our planet. But above all, it's composed of fourth-fifths nitrogen, the most abundant gas in Earth's atmosphere. The rest is made up of methane and other volatile hydrocarbons like butane and propane, along with traces of ammonium and a few other molecules.

This makeup is unique in the solar system and resembles much of what could be imagined of Earth's atmosphere in the earliest moments of existence, when the fundamental molecules of life would have been formed.

Of course, Titan is much too cold for life to evolve as it did on Earth. The surface temperature is –290°F (–180°C) and water exists as hard ice. Nevertheless, scientists are excited about studying Titan's original environment, hoping to get further information on the chemistry behind the origins of life.

Titan is one of the main objectives of the *Cassini-Huygens* spacecraft, a joint operation of NASA's *Cassini* probe, expected to enter orbit around Saturn in July 2004, and the European *Huygens* capsule, scheduled to descend into Titan's atmosphere and land on the surface of this mysterious satellite. What will *Huygens* discover under the halo of hydrocarbons? Astronomers imagine lakes or seas of methane with propane icebergs and ice continents traversed by rivers of methane. But will there be enough light? And will Saturn's majesty occasionally be visible through a gap in the clouds?

1

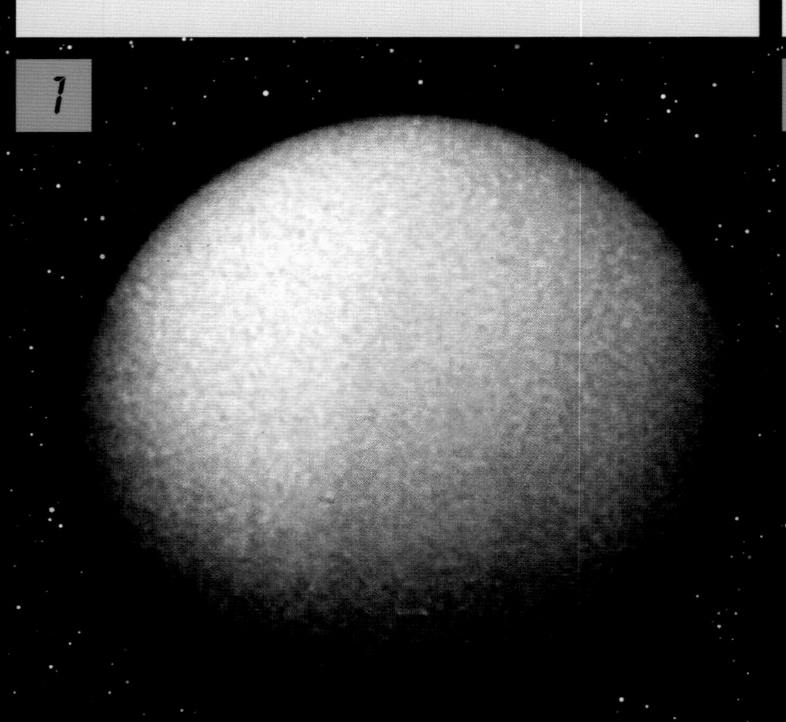

2

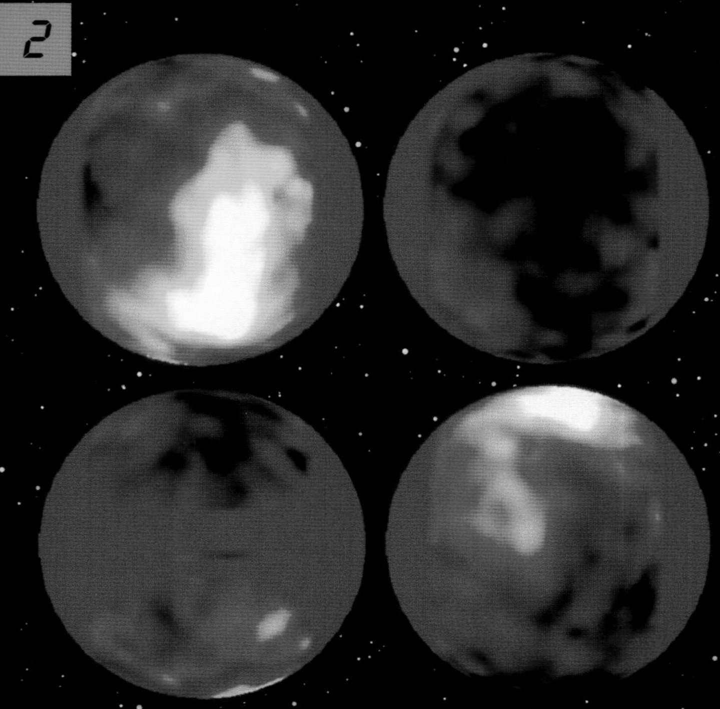

1 *TITAN DISCOVERED BY* VOYAGER *WITH ITS ORANGE-COLORED COVERING CREATED BY A HYDROCARBON FOG.*

2 *THE HUBBLE SPACE TELESCOPE PENETRATED TITAN'S CLOUDS TO SHOW A SURFACE WITH BRIGHT AREAS THAT COULD BE ICE CONTINENTS AND DARK PARTS THAT COULD BE METHANE OCEANS.*

3 *THE FOGGY COVER OVER THE POLAR AREAS ...*

4 *... COULD BE FORMED OF DROPLETS OF HYDROCARBONS AND AMMONIUM.*

5 HUYGENS *MAKING ITS DESCENT IN NOVEMBER 2004 INTO TITAN'S ATMOSPHERE: WILL THE LANDSCAPES RESEMBLE THOSE IMAGINED BY THE ASTRONOMERS ...*

6 *... WITH ICE CLIFFS AND METHANE LAKES? REALITY MAY BE EVEN MORE SURPRISING.*

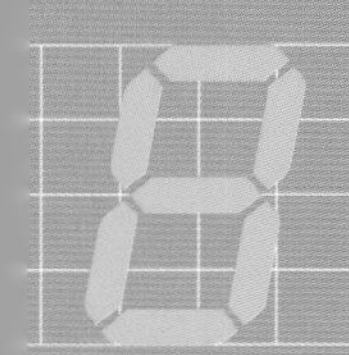

Bizarre, Bizarre Uranus

Twice as far from the Sun as Saturn and much less massive than it or Jupiter, Uranus is yet another of those balls of predominantly hydrogen gas that have come together in the icy reaches of the solar system, forced together at the whim of collisions of matter from multiple small planets of ice (or planetoids).

Despite its modest size (290 miles/470 km in diameter), one of the satellites of Uranus, Miranda, is without a doubt the best indication of that far-off time when titanic shocks controlled planetary formation. Miranda's relief is so irregular and so strange that it can only be the result of a succession of pieces smashing apart and coming together to finally gather into an impossible disorder. The other satellites of Uranus, including Titania, Oberon, Umbriel and Ariel, are less original; they're large chunks of rock and ice.

Uranus, on the other hand, has a very strange feature. Unlike Jupiter and Saturn where the atmospheres are brightly colored and divided into parallel areas at the equator, with occasional cyclonic formations like the famous Red Spot, Uranus has a uniformly blue face. Its rotational axis is perpendicular to the plane of its movement around the Sun, and its poles therefore point alternately to the Sun during the planet's year, which lasts 84 Earth years. Such a placement should create very distinct seasons, with a strong contrast between the two hemispheres but, on the contrary, the climate of Uranus is very homogeneous with a mean temperature of –365°F (–220°C) at the cloud tops. Another mystery is that the north magnetic pole is 60 degrees away from the geographic North Pole, sort of as if Earth's North Pole were in Cairo – bizarre, very bizarre!

Despite any interest, there is little reason for astronauts to make the trip to Uranus or its satellites someday. Robots will be sufficient to explore these far-off worlds, unless some economic reason causes people to go there. The atmosphere of Uranus, like that of Neptune, could be used to extract helium-3, the thermonuclear fuel that people expect to find on the Moon. The available stores of helium-3 on Uranus or Neptune are much larger, however; they would supply the energy needs of humanity for billions of years! Will tankers of helium-3 travel across the solar system one day? Nothing is impossible!

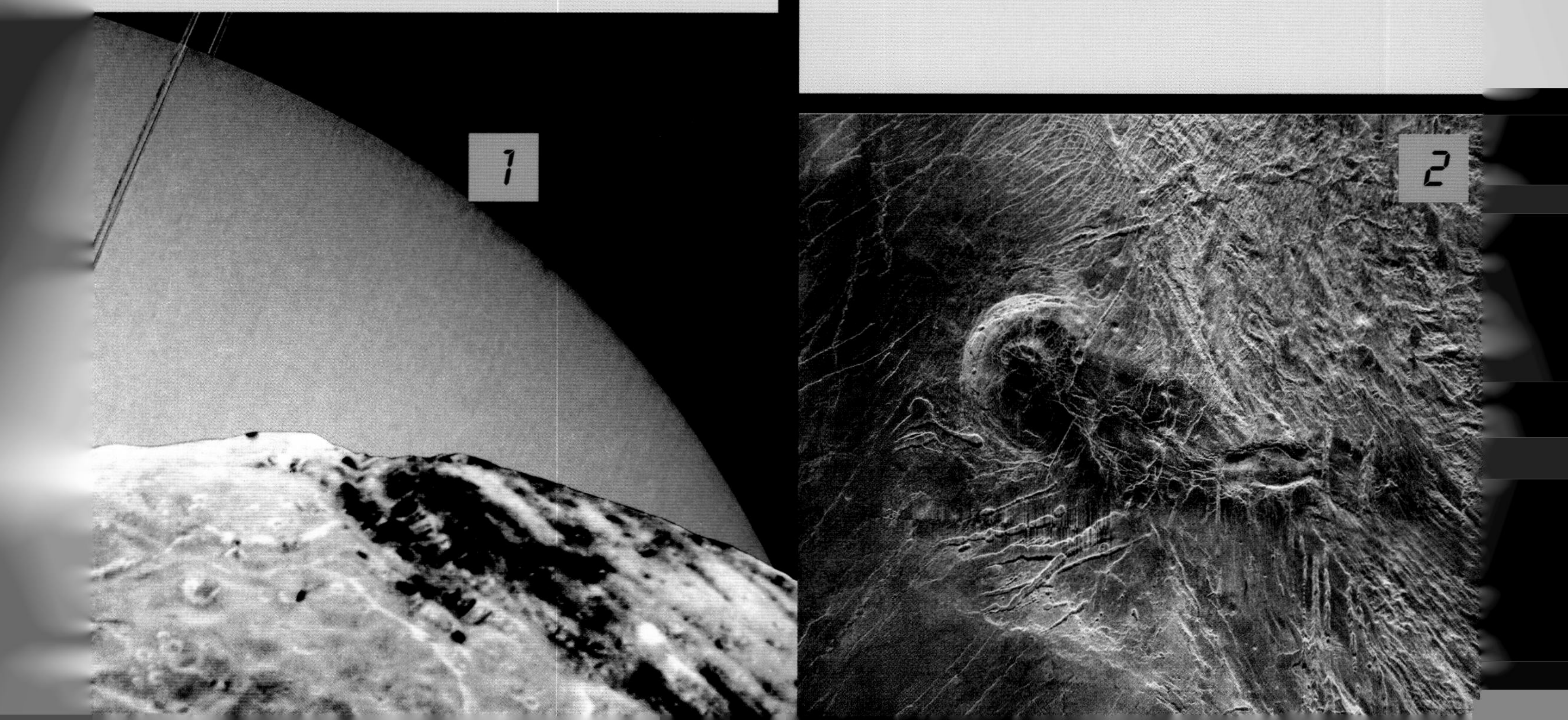

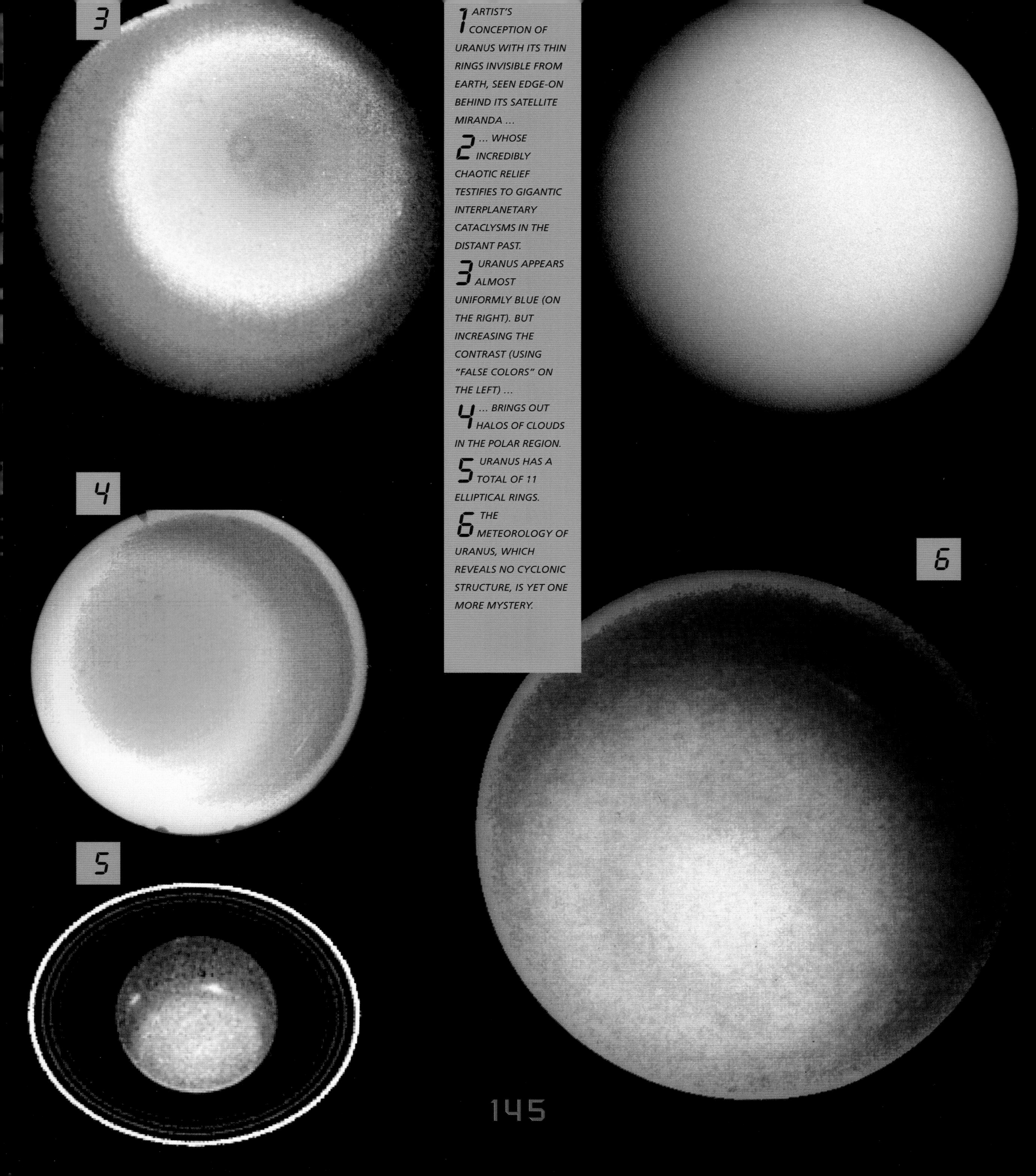

1 ARTIST'S CONCEPTION OF URANUS WITH ITS THIN RINGS INVISIBLE FROM EARTH, SEEN EDGE-ON BEHIND ITS SATELLITE MIRANDA ...

2 ... WHOSE INCREDIBLY CHAOTIC RELIEF TESTIFIES TO GIGANTIC INTERPLANETARY CATACLYSMS IN THE DISTANT PAST.

3 URANUS APPEARS ALMOST UNIFORMLY BLUE (ON THE RIGHT). BUT INCREASING THE CONTRAST (USING "FALSE COLORS" ON THE LEFT) ...

4 ... BRINGS OUT HALOS OF CLOUDS IN THE POLAR REGION.

5 URANUS HAS A TOTAL OF 11 ELLIPTICAL RINGS.

6 THE METEOROLOGY OF URANUS, WHICH REVEALS NO CYCLONIC STRUCTURE, IS YET ONE MORE MYSTERY.

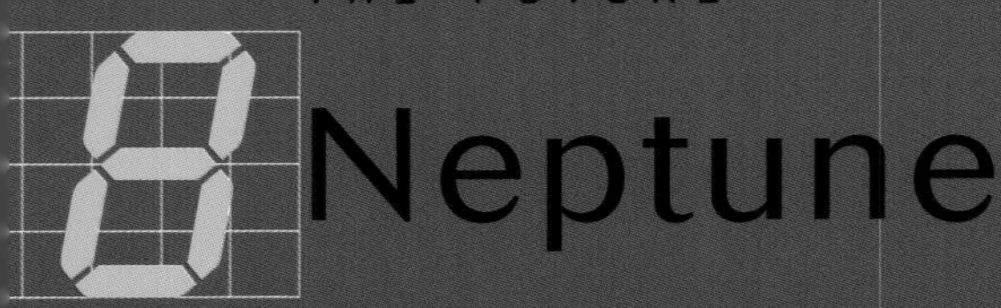

Neptune

Neptune is aptly named. It's the Jovian planet farthest from the Sun and is composed mainly of water. Is that the origin of its deep blue color? No, because Neptune's waters are shrouded by an atmosphere of helium and hydrogen, with a little methane, which absorbs red light and therefore reflects blue in return. The same phenomenon also explains the uniform blue of Uranus. However, even though it's the same size and color as its neighbor, Neptune is very different from the strange Uranus. It's more massive because of the large amount of water in its makeup, and it also has a much larger, rocky, metallic core. Its atmosphere is more active and includes a major cyclonic formation: the Great Dark Spot, a beautiful, dark blue oval bordered by white methane crystals. It also has a second dark spot, which is smaller and closer to the South Pole, and a third, rapidly moving tiny formation, called "Scooter."

Seen from Earth, Neptune is a single blue point whose light takes four hours to reach us. No surprise then that it wasn't until the *Voyager 2* spacecraft visited the planet in 1989 that its true features, its four thin rings and six of its 13 satellites could be uncovered. The largest of its satellites, Triton, was reputed to be the largest in the solar system. *Voyager 2* uncovered the truth: Triton is very bright because its icy surface is young, but it's smaller than the Moon and much smaller than Ganymede and Titan. That being said, it has its share of strange features: nitrogen geysers break through places in its ice surface, a "cryovolcanism" that's quite amazing at such low temperatures. Will Triton be humanity's last forward base at the edges of the solar system? Most likely not, because there are many other objects beyond Neptune's orbit, especially Pluto, which is the farthest planet from the Sun most of the time and seems quite different from Neptune and the other Jovian planets.

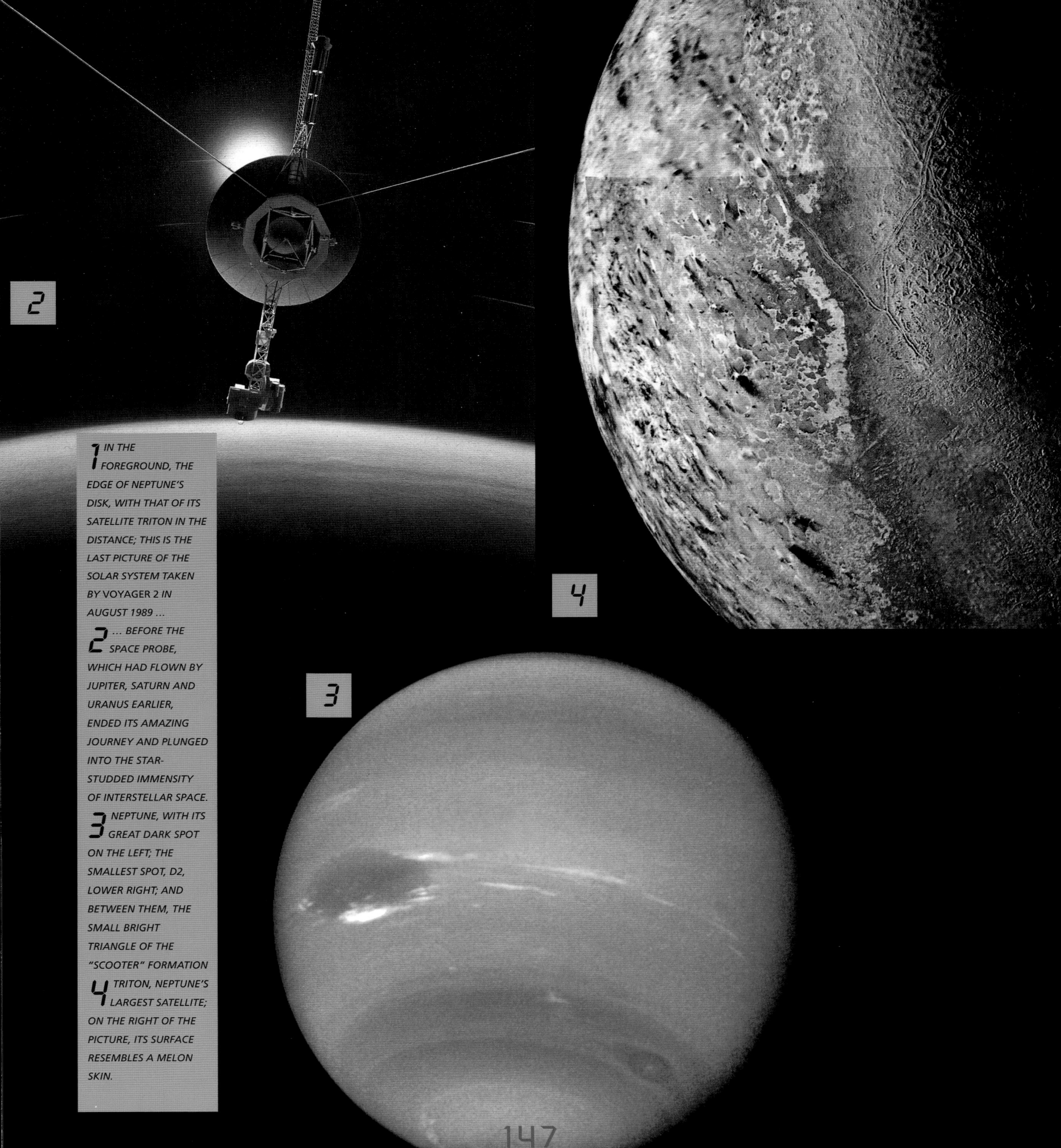

1 IN THE FOREGROUND, THE EDGE OF NEPTUNE'S DISK, WITH THAT OF ITS SATELLITE TRITON IN THE DISTANCE; THIS IS THE LAST PICTURE OF THE SOLAR SYSTEM TAKEN BY VOYAGER 2 IN AUGUST 1989 ...

2 ... BEFORE THE SPACE PROBE, WHICH HAD FLOWN BY JUPITER, SATURN AND URANUS EARLIER, ENDED ITS AMAZING JOURNEY AND PLUNGED INTO THE STAR-STUDDED IMMENSITY OF INTERSTELLAR SPACE.

3 NEPTUNE, WITH ITS GREAT DARK SPOT ON THE LEFT; THE SMALLEST SPOT, D2, LOWER RIGHT; AND BETWEEN THEM, THE SMALL BRIGHT TRIANGLE OF THE "SCOOTER" FORMATION

4 TRITON, NEPTUNE'S LARGEST SATELLITE; ON THE RIGHT OF THE PICTURE, ITS SURFACE RESEMBLES A MELON SKIN.

Pluto, Charon and the Kuiper Belt

Pluto is the ninth planet in the solar system, but its dimensions and nature make it more like the large icy satellites of the Jovian planets, like Triton more than Jupiter, Saturn, Uranus or Neptune. It's actually the smallest planet in the solar system at only 1,400 miles (2,300 km) in diameter, which explains why it wasn't discovered until 1930.

Its orbit is much longer than those of the other planets, varying in distance from the Sun between 2.7 billion and 4.6 billion miles (4.4 billion and 7.4 billion km), which occasionally brings it closer to the Sun than Neptune. This orbit is inclined 27 degrees from the ecliptic, close to where the other planets travel. Pluto has one satellite, Charon – seen for the first time in 1978 – only twice as small as it and very close to it, so much so that it could even be called a double planet. Besides their dimensions and very low temperatures (–375°F/–225°C), little is known of Pluto and Charon, which were not visited by any spacecraft in the 20th century. But a NASA probe should make up for that at the beginning of the 21st century.

Pluto is a very different planet that no longer belongs to the zone where the giant gas planets were formed, but rather to even more distant regions of the solar system, where planetoids of rock and ice exist in very large numbers, forming the Kuiper Belt that may stretch up to 9 billion miles (15 billion km) from the Sun. In November 2003, an object designated 2003 VB12 and nicknamed Sedna was discovered in an area three times as far from the Sun as Pluto. Sedna is estimated to be about 1,100 miles (1,770 km) in diameter. Scientists are still debating the "planetoid's" true origin and nature.

1

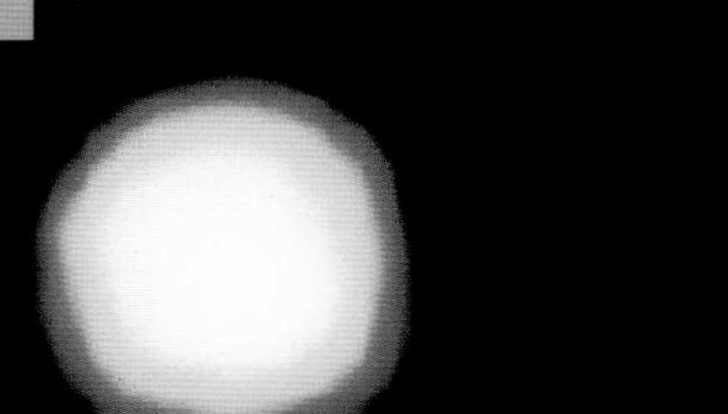

2

1 THE MOST ACCURATE MAP OF PLUTO AVAILABLE AT THE BEGINNING OF THE 21ST CENTURY, OBTAINED IN 1998 BY THE HUBBLE SPACE TELESCOPE …

2 … WHOSE FAINT OBJECT CAMERA ALSO PROVIDED IMAGES OF PLUTO AND ITS SATELLITE CHARON.

3 NOT ONLY DO PLUTO AND CHARON GUARD THE MYTHOLOGICAL GATES OF HELL, BUT ALSO THOSE OF THE KUIPER BELT, WHICH EXTENDS BEYOND NEPTUNE'S ORBIT AND INCLUDES TENS OF THOUSANDS OF ROCKY AND ICY OBJECTS.

4 BY 2010, A NASA PROBE IS EXPECTED TO FLY BY PLUTO AND CHARON, ILLUSTRATED HERE WITH THE SUN IN THE BACKGROUND, BEFORE EXAMINING OBJECTS IN THE KUIPER BELT.

3

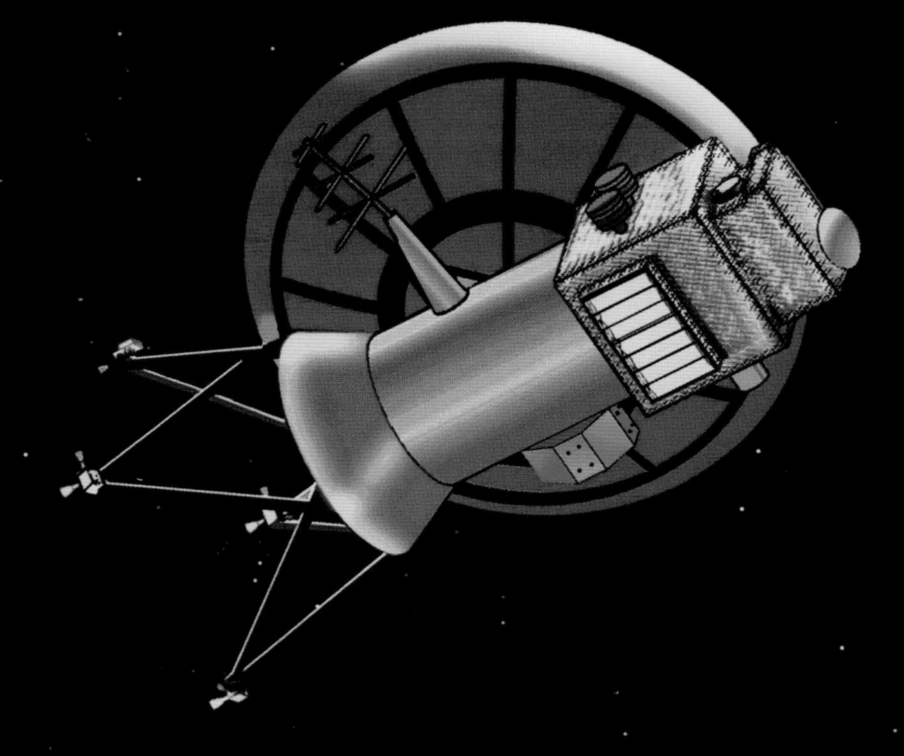

4

Comets and the Oort Cloud

Will people live on comets one day? The question seems absurd. Aren't comets short-lived bodies that burn up in the Sun's rays like moths attracted to light bulbs, providing a brief but spectacular show to Earth's captivated, yet concerned audience? Don't comets forewarn of terrible events to come?

In the life of a comet, the passage close to the Sun and the spectacular evaporation of its volatile matter, forming its bright head and long, diffuse tail, is just one exceptional episode. Chunks of rock dust and ice crystals (made up of water but also other elements such as ammonia, carbon dioxide, methane and so on), comets exist by the billions not only within the confines of the solar system, in the Kuiper Belt beyond Neptune, but also across a much wider area, the Oort Cloud, which encircles the solar system up to a distance of 9 trillion miles (15 trillion km). The Oort Cloud, like the Kuiper Belt, has existed since the early days of the formation of the solar system, some 4.5 billion years ago.

Occasionally, a comet's trajectory is disrupted by the attraction of a star, and it then begins the long journey which will, in thousands or millions of years, take it close to the Sun. Those that come from the Oort Cloud are usually "long-period comets" that leave for the endless reaches of space not to return for at least 200 years – or may never return. Those coming from the Kuiper Belt are often "short-period comets" that end up in regular orbits around the Sun, with periods under 200 years. The most famous one is Halley's Comet, which returns every 76 years and last passed by in 1986, greeted by an "armada" of space probes.

The great physicist and visionary Freeman Dyson feels that comets would be ideal for setting up space colonies. They have water, hydrocarbons and nitrogen compounds – that is, the atoms that create the molecules of life. Comets are not very big – most often a few miles to a few dozen miles in diameter, and perhaps a few hundred miles for some objects from the Kuiper Belt. But there are enormous numbers of them: about 10 trillion in the Oort Cloud, which could be home to populations thousands of times larger than humanity today!

But who would want to live in such far-flung regions of space where the Sun is only one star among many, on small, icy rocks as distant from each other as Earth is from Mars? The people of tomorrow might, however, be very different from us and seek the tranquility and independence of life on a comet. Freeman Dyson even imagined giant trees growing on comets with their roots pushing down to the comet's core and their leaves collecting the weak solar power.

1

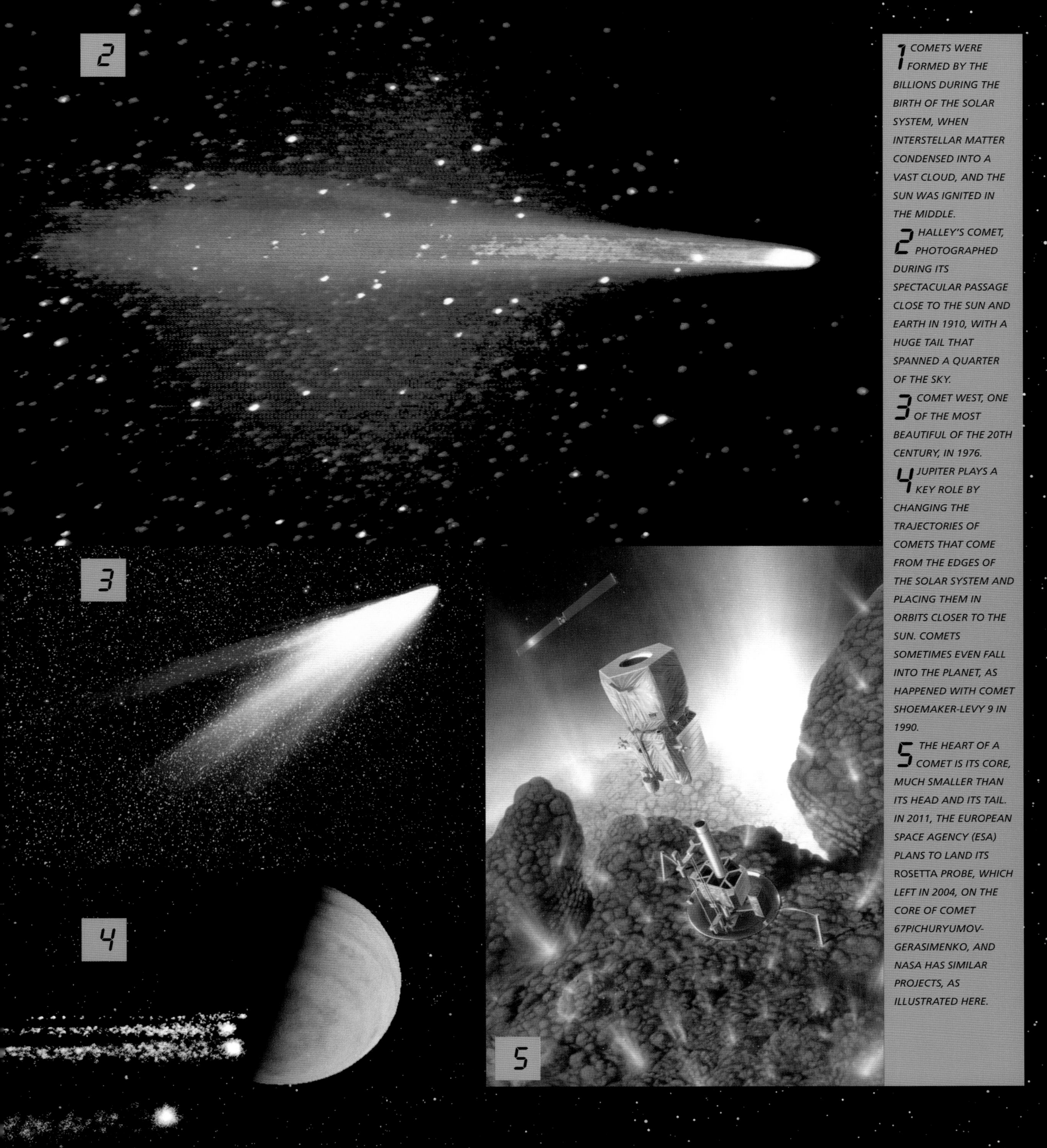

1 COMETS WERE FORMED BY THE BILLIONS DURING THE BIRTH OF THE SOLAR SYSTEM, WHEN INTERSTELLAR MATTER CONDENSED INTO A VAST CLOUD, AND THE SUN WAS IGNITED IN THE MIDDLE.

2 HALLEY'S COMET, PHOTOGRAPHED DURING ITS SPECTACULAR PASSAGE CLOSE TO THE SUN AND EARTH IN 1910, WITH A HUGE TAIL THAT SPANNED A QUARTER OF THE SKY.

3 COMET WEST, ONE OF THE MOST BEAUTIFUL OF THE 20TH CENTURY, IN 1976.

4 JUPITER PLAYS A KEY ROLE BY CHANGING THE TRAJECTORIES OF COMETS THAT COME FROM THE EDGES OF THE SOLAR SYSTEM AND PLACING THEM IN ORBITS CLOSER TO THE SUN. COMETS SOMETIMES EVEN FALL INTO THE PLANET, AS HAPPENED WITH COMET SHOEMAKER-LEVY 9 IN 1990.

5 THE HEART OF A COMET IS ITS CORE, MUCH SMALLER THAN ITS HEAD AND ITS TAIL. IN 2011, THE EUROPEAN SPACE AGENCY (ESA) PLANS TO LAND ITS ROSETTA *PROBE, WHICH LEFT IN 2004, ON THE CORE OF COMET 67PICHURYUMOV-GERASIMENKO, AND NASA HAS SIMILAR PROJECTS, AS ILLUSTRATED HERE.*

Manned Flights to Distant Planets

After the trip to Mars, no doubt trips will follow to Europa, then Titan and, even later, the satellites of Uranus, then Triton, Pluto and Charon, objects in the Kuiper Belt, comets in the Oort Cloud – a program of exploration and, depending on circumstances, exploitation and colonization for the third millennium! In truth, it would take a lot of time to develop new technology allowing rapid travel in the vast spaces of the solar system and the cosmic immensity stretching beyond Neptune. The nuclear plasma VASIMR rocket on the *Tsiolkovski* enables Earth-Mars trips in a few months. But distances increase rapidly beyond the red planet. Jupiter is 465 million miles (750 million km) from the Sun, compared to an average of 129 million miles (207 million km) for Mars; Saturn is 930 million miles (1.5 billion km) from the Sun, Uranus 1.9 billion miles (3 billion km) and Neptune 2.8 billion miles (4.5 billion km). VASIMR technology could be used to reach Jupiter, and thus the mysterious Europa, in under a year, which is quite acceptable for an initial mission of exploration. Consequently, the first trip to the Jovian world might be close to 2050, but setting up a base on Europa with regular return flights, and especially venturing farther into the solar system, demands new technology beyond the bounds of nuclear physics.

Fusion of the nuclei of light atoms – that is, the power source that drives the Sun (and thermonuclear bombs!) – would doubtlessly be the best solution at first, but that's still far away. Physicists and engineers have been studying the control of nuclear fusion since the 1950s, but no functional reactor exists after half a century of effort. A fusion-powered rocket motor will not see the light of day until the last decades of the 21st century, so trips to bodies ranging from Titan to Pluto will certainly not take place before 2080 to 2100. Even then, new technologies will be needed to go further and penetrate the Kuiper Belt.

The solution at that time will unquestionably be matter-antimatter annihilation, which will release the enormous amounts of energy contained in matter, using Einstein's famous formula $E=mc^2$. A few grams of matter and antimatter would be all that is needed to provide thrust for a quick trip to Pluto! But formidable obstacles lie in the path of this technology. It would first have to be possible to manufacture and store large quantities (grams) of antimatter. Then, the incredibly violent matter-antimatter reactions would have to be controlled. Physicists and engineers will undoubtedly develop the antimatter rocket motor in the first half of the 22nd century. Trips in the solar system will be much easier then. Mars will be reachable in a few days, Pluto in a few weeks, and the heart of the Kuiper Belt in a few months. The age of space colonization will have truly begun!

1 A MANNED SPACECRAFT FLIES CLOSE BY IO IN THE JUPITER SYSTEM – AN IMAGE FOR THE END OF THE 21ST CENTURY.

7

8

Does Life Originate in Space?

1

2

On Mars during the *Tsiolkovski* mission, as later on Europa and Titan, it's the search for their origins that motivates humanity's drive to explore. Space probes and astronauts are sent to seek out information to explain the appearance and development of life on Earth. The discovery of a life form elsewhere in the solar system would be particularly significant. If that life form had the same molecular basis as terrestrial life (that is, proteins and ribonucleic acids) there would be a strong likelihood that all life in the universe resembles the terrestrial model and requires, especially, water and warm temperatures. If, on the other hand, a completely different type of life form were found, it would indicate that the phenomenon of life may be more diverse and manifest in environments very different from those on Earth. But would we be able to recognize a life so different from the only one we know – ours? That's not a sure thing, as the ambiguous conclusions of Mars exploration have shown. It's not even certain that terrestrial life emerged on Earth!

At the beginning of the century, the chemist Svante Arrhenius proposed an original, heretical theory of life on Earth. "Panspermia," single-celled organisms in the form of spores capable of surviving in the interstellar environment for millions of years, might have seeded our planet. It's a fact that spores can survive space conditions. It's also a fact that pieces of planets can be ripped away during collisions and projected into space, such as, for example, the Martian meteorites found on Earth. Could the first living organisms thus have arrived on Earth from Mars, from another body in the solar system, or even from another star despite the

3

4

5

1 NEBULAS, ENORMOUS CLOUDS OF INTERSTELLAR GAS, ARE STAR NURSERIES WHERE MYRIAD SUNS ARE BORN.

2 THE MAGNIFICENT WHIRLPOOL SPIRAL GALAXY, CONTAINING SOME 100 BILLION STARS, HAS A DIAMETER OF SOME 100,000 LIGHT-YEARS …

3 … AND CERTAINLY BEARS A STRONG RESEMBLANCE TO OUR OWN GALAXY, THE MILKY WAY, WHERE THE SUN IS AN ORDINARY STAR, LOCATED SOME 27,000 LIGHT-YEARS FROM ITS CENTER AT THE EDGE OF ONE OF ITS SPIRAL ARMS.

4 HOW MANY OF THESE FORMING STARS WILL ONE DAY RESULT IN THE EMERGENCE OF A LIFE FORM?

5 THE FAMOUS HORSEHEAD NEBULA.

immense interstellar distances? It may seem very unlikely, but no theory can be excluded.

Even if terrestrial life did not come from space, another question arises: Are the organic molecules that are certainly behind the origin of the evolution toward life not of cosmic origin? It's a fact that organic molecules exist in nebulas where stars are born. It's also known that the comets forming in those nebulas also contain organic molecules, and that oceans on Earth probably originated during frequent bombardments of the planet by comets. So isn't it true that molecules from space played a key role in the origins of life on Earth?

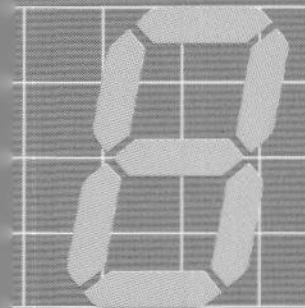

The Search for Extrasolar Planets

A life form similar to that on Earth can only appear on a planet where water exists or existed in a liquid state for hundreds of millions, or even billions of years. In the solar system, only Mars and perhaps Europa satisfy this criterion. But are there other "habitable" planets around other stars? Astronomy can't answer that question at the beginning of the 21st century; it doesn't yet have the instruments for detecting the presence of such planets (that is, planets that are small and rocky and receiving just the right amount of solar energy so a climate favorable to life can exist). On the other hand, other much more massive planets the size of Jupiter or even larger can be detected around stars. These "extrasolar planets," or exoplanets, are not directly visible, but their presence around a star is indicated by slight fluctuations in its trajectory. This "astrometric" method was used to detect close to a hundred in the closing decades of the 20th century.

Do stars accompanied by massive exoplanets have stellar systems comparable to the solar system, with numerous planets, some of which might resemble Earth? Astronomers tend to think so. In the nebula that condenses around a forming star, many objects must amass to give birth to many planets. To be sure, more accurate "astrometric" observatories, such as the ESA's *Gaia* craft (scheduled for about 2010) are needed to detect planets with Earth's mass. Another technology must be available to identify the slight drop in the light from a star when a planet crosses in front of it. It will be used by the French Corot Observatory (in orbit in 2006). NASA's Kepler Observatory (planned for 2007) will also provide information on the presence and composition of an atmosphere. Later, in the 2010s, even more productive space observatories, such as the ESA's *Darwin*, should provide images of planets the size of Earth around stars at distances of up to 30 light-years from the Sun.

A large quantity of oxygen observed in the atmosphere of such a planet would lead people to believe that some form of life exists. This would be an extraordinary discovery, even if it were completely impossible to go and see what that life form looked like up close.

1 HOW MANY STARS ARE SURROUNDED BY STELLAR SYSTEMS? MASSIVE PLANETS HAVE BEEN DETECTED AROUND SOME 100 STARS.

2 ARTIST'S CONCEPTION OF A LARGE PLANET THE SIZE OF JUPITER, DETECTED AROUND STAR HD209458, 150 LIGHT-YEARS FROM THE SUN. IT'S THE FIRST EXOPLANET WHOSE ATMOSPHERE WAS STUDIED BY THE HUBBLE SPACE TELESCOPE IN 2001.

3 THE "NEXT GENERATION SPACE TELESCOPE" (NGST), WHICH WILL SUCCEED HUBBLE, WILL BE PART OF A SERIES OF SPACE OBSERVATORIES FOR DETECTING AND STUDYING EXOPLANETS THE SIZE OF EARTH.

Are We Alone in the Universe?

Finding other living organisms on Mars or Europa and detecting exoplanets around other stars where life forms surely exist would show that the phenomenon of life pervades the universe and that Earth is not unique. But on our planet, the emergence of life was just one step in the long sequence of events that led to humanity and civilization – the arrival of multicelled organisms, the development of plants, fish leaving the oceans, the emergence of mammals and the appearance of humans, culminating in the species *Homo sapiens*; and from there consciousness, intelligence, language, cave paintings, funeral rites, livestock and agriculture, writing, science and technology ... It's a long road; 3.8 billion years have passed since the first living cells appeared in the oceans! Have other planets been the site of similar developments? Do intelligent beings exist elsewhere in the universe? This question is even broader than that of the existence of other life forms. Nothing really proves that all life evolves according to the terrestrial model into increasingly complex beings capable of one day being able to develop the means to understand the universe and ask questions about "the silence of infinite space." Is Earth civilization not the result of a series of extraordinary coincidences? Are we not alone in the universe?

Radio astronomers studying the radio waves emitted by celestial bodies have proposed a method for trying to answer these basic questions: Seek out signals that might have been sent out into space by extraterrestrial civilizations, intended for beings like us who have just learned to use radio technology. If other intelligent beings exist, their civilizations

3

2

1 HOW MANY CIVILIZATIONS CREATED BY INTELLIGENT BEINGS ARE THERE IN THE UNIVERSE? OUR GALAXY ALONE, THE MILKY WAY, HAS NO FEWER THAN A HUNDRED BILLION STARS …

2 … AND THE UNIVERSE CONTAINS TENS OF BILLIONS OF GALAXIES.

3 MANY RADIO TELESCOPES, LIKE THE ONE IN NANÇAY, FRANCE, ARE LISTENING TO THE STARS, LOOKING FOR SIGNALS THAT MIGHT COME FROM EXTRATERRESTRIALS.

might date back millions, even billions, of years – the universe is 15 billion years old and the solar system 4.5 billion. They could have attained a level of technology we cannot even imagine, but may still want to contact "emerging" civilizations like ours. These ideas led to the creation of an international program called SETI (Search for Extra-Terrestrial Intelligence). Radio telescopes focus on the sky, listening for signals that might be of artificial origin, sent out by civilizations far away in space and time. At the beginning of the 21st century, no signal so far seems to have been discovered coming from an artificial source. But would we be able to recognize such a signal? Would we be able to understand different beings who are much more advanced than us?

Will We Travel to the Stars?

People will no doubt be exploring the entire solar system by the early centuries of the third millennium, including the icy planetoids in the immense Kuiper Belt and areas close to the gigantic Oort Cloud, up to some tens of billions of miles from the Sun. Humanity will establish itself on Mars, Europa, Titan and many other celestial bodies, creating a civilization that will become increasingly interplanetary. We will then have reached the edge of a monstrous gulf separating the solar system from the nearest stars, so vast that it can no longer be measured in billions of miles but in light-years; the closest star, Alpha Centauri, is 4.3 light-years away – that is, 25 million billion miles (40 million billion km).

Will we be able to cross this gulf one day? Creative physicists and engineers have started thinking about the technologies that would make it possible to traverse interstellar space and reach other stars. Used on a massive scale, thermonuclear fusion, which will certainly be used for trips in the solar system in years to come, may make it possible to build spaceships that could reach speeds that are a small percentage of light speed: around 6,200 miles per second (10,000 km/s), 300 times the speed of Earth around the Sun! But even at such speeds, reaching Alpha Centauri would take more than a century, and that's not counting the time it would take to slow the ship down as it approached the star.

Other technologies have been considered, such as immense solar sails propelled by rays of light with the power of a star. But there is no chance any of them will be mastered in the third millennium. Further, traveling to Alpha Centauri would be just a tiny step in interstellar space. The Milky Way (the galaxy to which our Sun belongs) measures no less than 100,000 light-years in diameter, and it's separated from the closest galaxy by several hundreds of thousands of light-years.

The universe is too vast to cross for interstellar spacecraft built according to known laws of physics. But can't physics evolve? Can new laws enabling unimaginable voyages in space-time be found? Didn't the 20th century rewrite the physics of earlier centuries with relativity and quantum mechanics? The future of science may reveal many surprises, but it's impossible to know which ones. This area of speculation is the realm of science fiction, and that's best left to the excellent authors of the genre who allow us to dream, and who may actually describe a future where people will, in some as yet unknown way, travel the universe. One thing is certain: Without a scientific revolution, interstellar space will remain unattainable.

V
The Universe

Are the conquest of Mars and the spread of humanity into the solar system the prelude to the conquest of the universe? Science fiction is full of heroes traveling across intergalactic space and civilizations extending thousands and millions of light-years into space. Will humanity one day make the dreams of science fiction a reality? Have other beings elsewhere in the universe already experienced such fantastic adventures?

Reaching the closest stars, less than 10 light-years distant, would be difficult with the technology we can imagine today. Moreover, the existence of other civilizations in the universe is a question without an answer, even after the Mars 1 mission, which seemed to confirm that life has every chance of emerging as long as the right conditions come together. Traveling across interstellar space and meeting other "intelligent" beings are still unrealistic dreams. Will they come true one day? Nobody can seriously answer that question. On the other hand, it is possible to take measurements of this universe where we live, but are only an infinitesimal part of – a species that appeared only a few million years ago on a small planet, circling a moderate star, in an average-sized galaxy, while there are tens of billions of galaxies, each containing hundreds of billions of stars....

The progress of astronomy and the conquest of space allow us to measure the immensity of the universe, in time and space – in the approximately 15 billion years that have passed since the Big Bang, and in distances that extend out to 15 billion light-years. They also allow us to understand the extraordinary violence and fantastic complexity of the phenomena taking place there. Two space telescopes play a pivotal role in studying them: Hubble, which photographs the sky in the visible light-range, and Chandra, which operates with x-rays. Their pictures have opened up an exceptional window onto the beauties of the universe surrounding us.

1 AT 11 BILLION LIGHT-YEARS, TWO QUASARS OBSERVED BY CHANDRA – TWO "LIGHTHOUSES" LIGHTING UP THE EARLIEST AGES OF THE UNIVERSE, WHEN THE FIRST GALAXIES AND FIRST STARS WERE BORN A FEW BILLION YEARS AFTER THE BIG BANG.

2 CHANDRA, A MAGNIFICENT SPACE TELESCOPE BUILT TO PENETRATE THE SECRETS OF THE UNIVERSE, AND LAUNCHED IN 1997, RECORDS HIGH-ENERGY X-RAYS EMITTED DURING EXTREMELY VIOLENT COSMIC PHENOMENA.

3 THE HCG62 GROUP OF GALAXIES IN THE CONSTELLATION VIRGO, OBSERVED BY CHANDRA. AFTER FORMING IN THE EARLIEST TIMES OF THE UNIVERSE, THE GALAXIES CAME TOGETHER IN GROUPS, SOME OF WHICH ASSEMBLED TO FORM LARGER CLUSTERS. THE MILKY WAY BELONGS TO THE LOCAL GROUP WITH THE ANDROMEDA NEBULA AND MAGELLANIC CLOUDS.

4 THE ABELL 2142 GALAXY CLUSTER IS ONE OF THE MOST MASSIVE OBJECTS IN THE UNIVERSE: WITH A DIAMETER OF SIX MILLION LIGHT-YEARS, IT CONTAINS HUNDREDS OF GALAXIES AND IS THE RESULT OF A COLLISION OF TWO SMALLER CLUSTERS WHERE THE SHOCK WAVE IS REPRESENTED BY THE BLUISH HALO.

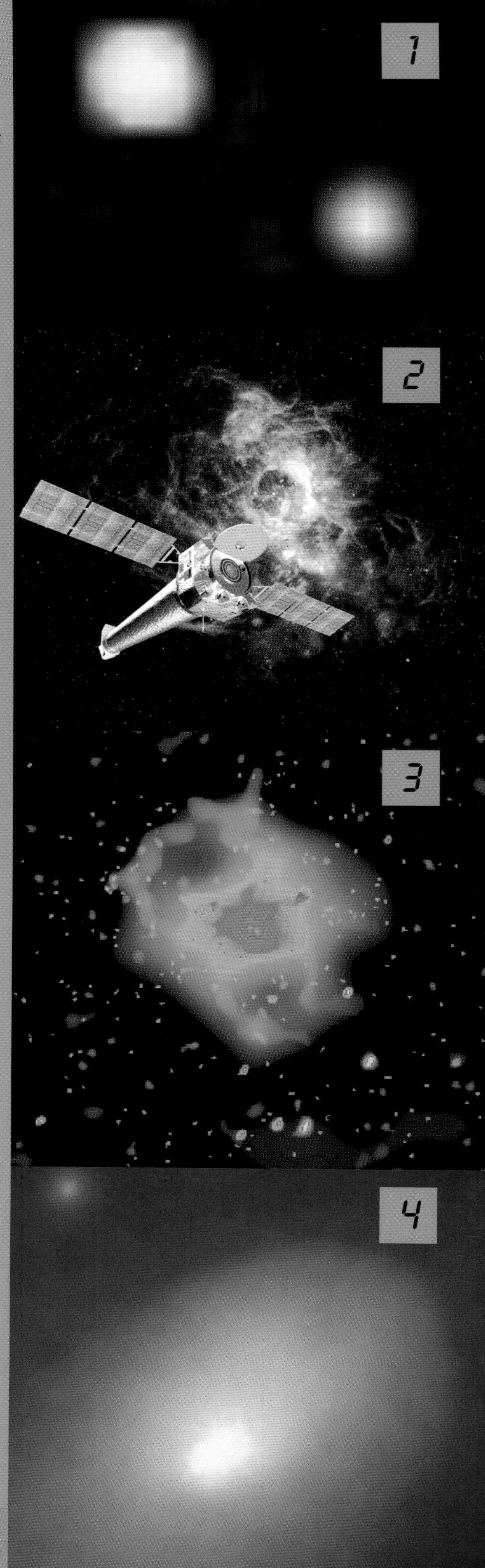

Most theories about life in the universe are based on the same assumption: A life form can only emerge on a planet like Earth, or Mars – in other words, around a star. But how long have stars existed? Findings obtained by Hubble seem to show that the stars formed much earlier than was thought, even before the great elliptical or spiral galaxies, like our Milky Way, appeared. Stars might have existed only a billion years after the Big Bang. This possibility seems to be confirmed by other Hubble images, apparently showing stars at the end of their life, white dwarves, which would be 12 billion to 13 billion years old. Of course, the first stars could not have been surrounded by planets like Earth; heavy chemical elements, such as those that make up our planet and us, did not yet exist. They formed in the heart of these first generations of stars through the process of nucleosynthesis. But stars much more massive than the Sun had short lives and ejected elements into space that could be used to form telluric planets. So possible havens for life have doubtlessly existed for over 10 billion years, while our Sun has been shining for only 4.5 billion years. What ever became of such ancient lives or "intelligent" beings?

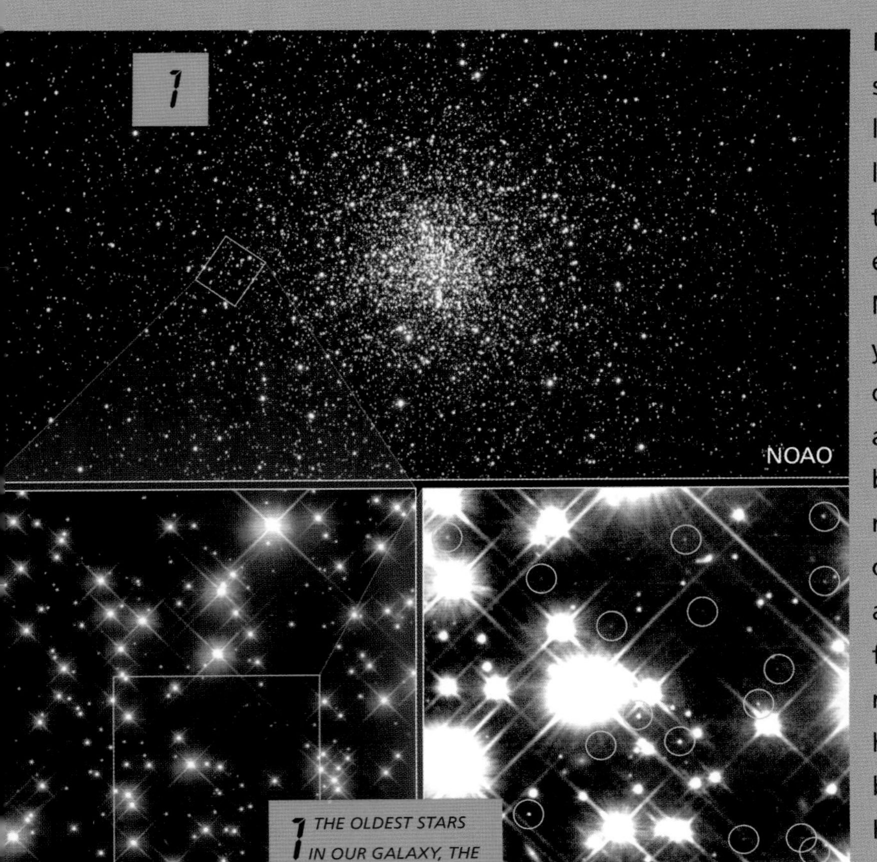

1 THE OLDEST STARS IN OUR GALAXY, THE MILKY WAY, PHOTOGRAPHED BY THE HUBBLE SPACE TELESCOPE. THEY ARE WHITE DWARVES 12 BILLION TO 13 BILLION YEARS OLD IN THE GLOBULAR CLUSTER M4, 7,000 LIGHT-YEARS FROM THE SUN.

2 INFORMATION PROVIDED BY HUBBLE HAS LED SOME ASTRONOMERS TO THINK THAT STARS FORMED VERY EARLY IN THE LIFE OF THE UNIVERSE, BEFORE THE GREAT ELLIPTICAL AND SPIRAL GALAXIES. THIS ARTIST'S CONCEPTION ILLUSTRATES THE FIREWORKS OF MASSIVE STARS FORMING AND SELF-DESTRUCTING IN SPECTACULAR SUPERNOVA EXPLOSIONS.

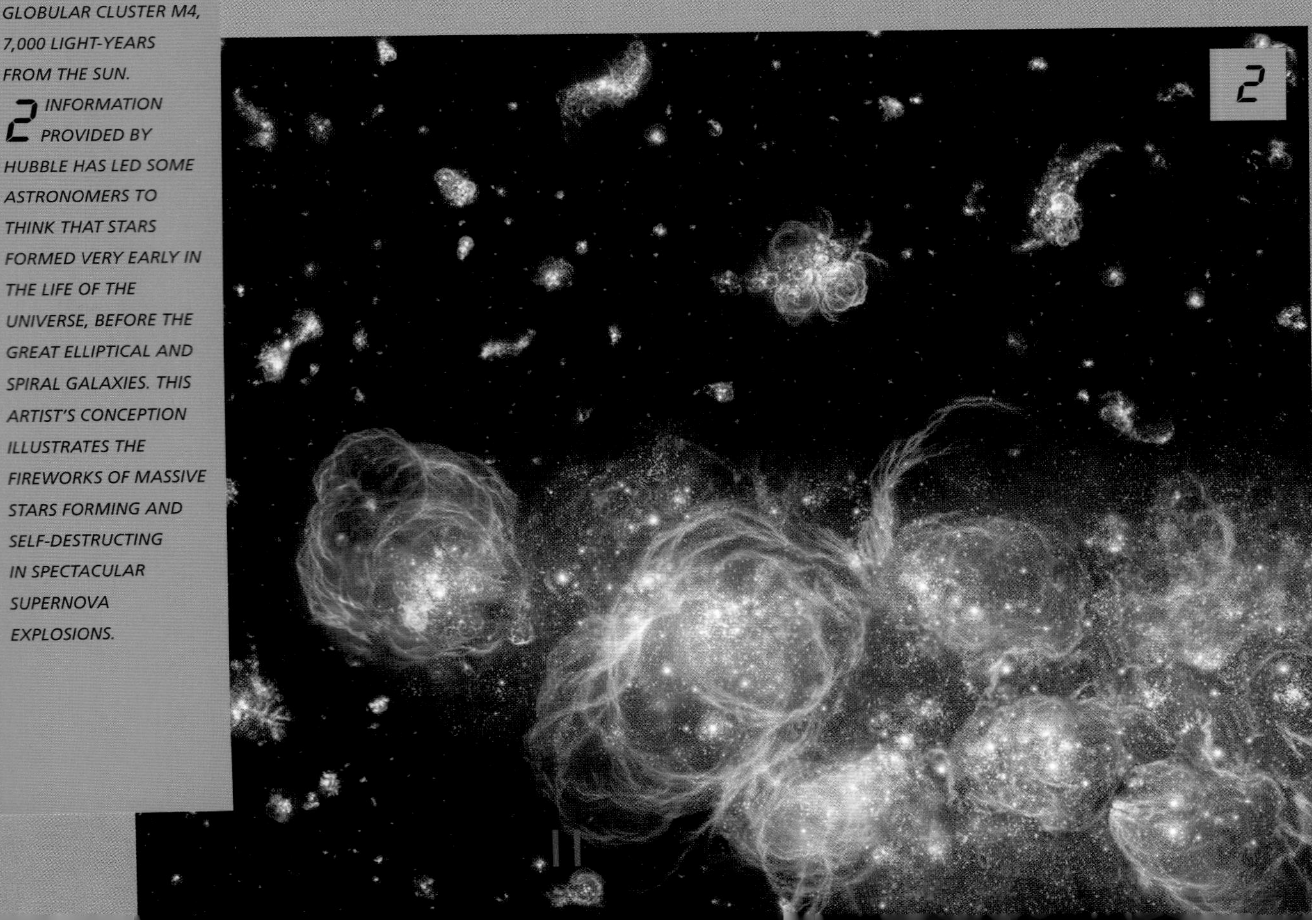

1

Sta
burn
combu
which g
caused th
collapse u
mass in catacly
and become sup
with moderate a
mass, such as the Su
a long time, so life
chance to evolve. The Su
shine for about another
billion years, but other mo
massive stars have much
shorter lives, which can be
counted in only millions of
years. Depending on the
situation, exploding stars give
birth to various types of
"exotic" space objects; they
become neutron stars, tiny
clusters of incredibly dense
iron cores that manifest in the
form of regular emissions of
radio waves (which is why
they're called pulsars), or are
even transformed into black
holes, points of infinite
gravity that suck in all matter
within range. Their existence
was postulated by Stephen
Hawking and Roger Penrose,
and the latest astronomical
observations seem to confirm
this extraordinary hypothesis.

1 ETA CARINAE IS THE BRIGHTEST STAR IN THE MILKY WAY; IT RADIATES ENERGY AT FIVE MILLION TIMES THE INTENSITY OF THE SUN. LOCATED 7,000 LIGHT-YEARS AWAY, IT'S UNSTABLE AND COULD EXPLODE AND BECOME A SUPERNOVA AT ANY TIME. THERE'S NO THREAT TO LIFE ON EARTH, BUT THE SPECTACLE WOULD BE IMPRESSIVE.

2 THIS EXPANDING CLOUD OF GAS, PHOTOGRAPHED BY CHANDRA, HAS A DIAMETER OF 10 LIGHT-YEARS AND A TEMPERATURE OF 50 MILLION DEGREES. IT WAS PRODUCED BY THE EXPLOSION OF A SUPERNOVA ONLY THREE CENTURIES AGO IN THE CONSTELLATION CASSIOPEIA.

n
per
ourse
comet
cause im
Earth, as th
occasions in t

VII

V

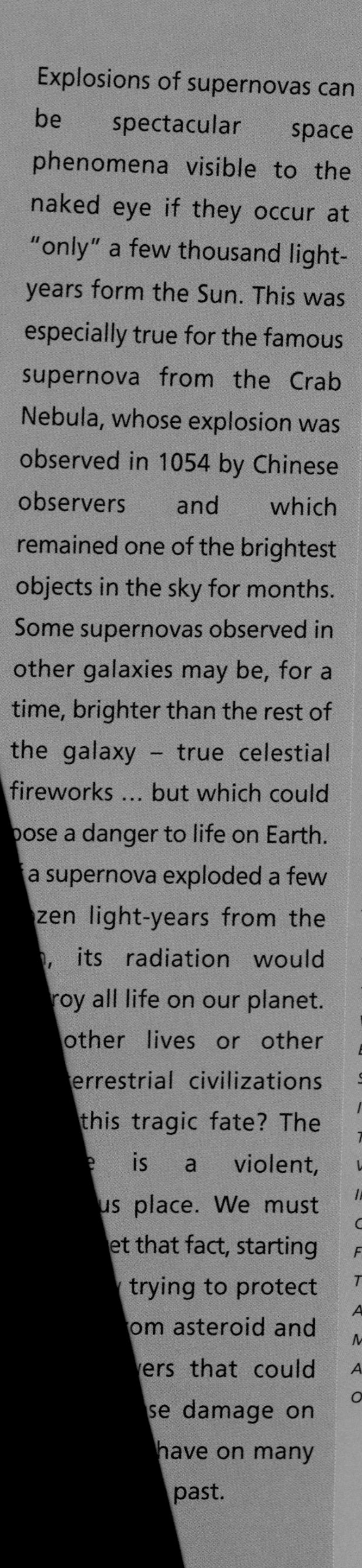

Explosions of supernovas can be spectacular space phenomena visible to the naked eye if they occur at "only" a few thousand light-years form the Sun. This was especially true for the famous supernova from the Crab Nebula, whose explosion was observed in 1054 by Chinese observers and which remained one of the brightest objects in the sky for months. Some supernovas observed in other galaxies may be, for a time, brighter than the rest of the galaxy – true celestial fireworks ... but which could pose a danger to life on Earth. a supernova exploded a few zen light-years from the , its radiation would roy all life on our planet. other lives or other errestrial civilizations this tragic fate? The e is a violent, us place. We must et that fact, starting trying to protect om asteroid and ers that could se damage on have on many past.

1 THE CRAB NEBULA IS THE REMNANT OF THE MOST FAMOUS SPACE EXPLOSION IN HISTORY: THAT OF THE SUPERNOVA IN 1054, SEEN BY CHINESE OBSERVERS. IT'S 6,000 LIGHT-YEARS FROM THE SUN. CHANDRA REVEALS ITS STRUCTURE HERE. AT THE CENTER IS A PULSAR, A RAPIDLY SPINNING NEUTRON STAR, THAT EMITS WAVES, A LITTLE LIKE A LIGHTHOUSE, AT A RATE OF 30 PULSES A SECOND.

2 WHEN IT EXPLODED APPROXIMATELY 10,000 YEARS AGO, THE SUPERNOVA IN THE CONSTELLATION VELA CREATED AN EXTRAORDINARY SPECTACLE IN THE SKY; IT NO DOUBT GREW 50 TIMES BRIGHTER THAN VENUS. THIS CHANDRA IMAGE SHOWS THE GAS CLOUD THAT RESULTED FROM THIS EXPLOSION TODAY, EXPANDING AT A SPEED OF 250,000 MPH (400,000 KM/H) AND WITH A DIAMETER OF 100 LIGHT-YEARS

It took millions of years for humanity to occupy the lands that emerged on our planet, to master the seas, to conquer the highest peaks, to reach for the sky. At the dawn of the third millennium, the road to the stars is finally opening up to the children of Earth. Borne by their dreams, they are now ready to challenge the universe, to conquer the infinite. Will they one day contemplate space, as we do here, from another Earth, circling another Sun, in another galaxy? A great adventure is beginning, and it will last as long as there are people to live it ...

Scientific Appendices

The Sun

Diameter: 866,000 miles (1,393,000 km)
Average distance from Earth: 93,000,000 miles (150,000,000 km)
Period of revolution: 225 million years around the axis of the galaxy
Period of rotation (days): -26 (equator)
-36 (poles)
Orbital speed: that of the galaxy
Main atmospheric components (C = clouds): H (hydrogen) + He (helium)
Temperature: exterior (photosphere): 10,800°F (6,000°C)
interior (center): 25,200,000°F (14,000,000°C)
Gravity: 273 times that of Earth
Mass: 332,000 times that of Earth
Volume: 1,300,000 times that of Earth
Magnetic field: extremely powerful
Known satellites: the planets and their satellites

Mercury

Diameter: 3,030 miles (4,880 km)
Average distance from Sun: 35,980,000 miles (57,900,000 km)
Minimal distance from Earth: 49,000,000 miles (79,000,000 km)
Period of revolution (years): 0.24 (88 days)
Period of rotation (days): 58.6
Average orbital speed: 2.9 miles/second (4.7 km/s)
Inclination of axis at poles: 0°
Inclination of orbit from ecliptic: 7°
Main atmospheric components (C = clouds): no atmosphere
Average surface temperature: day 660°F (350°C), night –275°F (–170°C)
Gravity: 0.38 (Earth = 1)
Mass: 3.3022 x 1,023 kg (Earth = 1)
Volume: 0.056 (Earth = 1)
Magnetic field: weak
Known satellites: 0

Venus

Diameter: 7,521 miles (12,104 km)
Average distance from Sun: 67,230,000 miles (108,200,000 km)
Minimal distance from Earth: 25,500,000 miles (41,000,000 km)
Period of revolution (years): 0.61 (224.7 days)
Period of rotation (days): 243
Average orbital speed: 22 miles/second (35 km/s)
Inclination of axis at poles: 2°
Inclination of orbit from ecliptic: 3.4°
Main atmospheric components (C = clouds): CO2 carbon 90% (+ sulfuric and hydrochloric acid), C++
Average surface temperature day/night: 895°F (480°C)
Gravity: 0.91 (Earth = 1)
Mass: 4.8689 x 1,024 kg (Earth = 1)
Volume: 0.88 (Earth = 1)
Known satellites: 0

Earth

Diameter: 7,926 miles (12,756 km)
Average distance from Sun: 92,960,000 miles (149,600,000 km)
Period of revolution (years): 1 (365 days)
Period of rotation (days): 1 (23 h 56′)
Average orbital speed: 18.5 miles/second (29.8 km/s)
Inclination of axis at poles: 23° 44′
Inclination of orbit from ecliptic: 0°
Main atmospheric components (C = clouds): oxygen, C
Average surface temperature: day 72°F (22°C), night 59°F (15°C)
Gravity: 1
Mass: 1 (5,975 x 1,024 kg)
Volume: 1
Magnetic field: powerful
Known satellites: 1 (Moon)

Mars

Diameter: 4,222 miles (6,794 km)
Average distance from Sun: 141,295,000 miles (227,392,000 km)
Minimal distance from Earth: 35,000,000 miles (56,000,000 km)
Period of revolution (years): 1.88 (687 days)
Period of rotation (days): 1.03 (24 h 37′)
Average orbital speed: 15 miles/second (24.1 km/s)
Inclination of axis at poles: 24° 46′
Inclination of orbit from ecliptic: 1° 9′
Main atmospheric components (C = clouds): CO2 (95%)/Ar, C
Average surface temperature: day –22°F (–30°C)
Gravity: 0.38 (Earth = 1)
Mass: 6.4191 x 1,023 kg (Earth = 1)
Volume: 0.15 (Earth = 1)
Magnetic field: weak
Known satellites: 2 (Phobos and Deimos)

Jupiter and its Satellites

Jupiter

Diameter: 88,700 miles (142,800 km)
Average distance from Sun: 483,600,000 miles (778,300,000 km)
Minimal distance from Earth: 362,000,000 miles (583,000,000 km)
Period of revolution (years): 11.86
Period of rotation (days): 0.41 (9 h 50′)
Average orbital speed: 8.1 miles/second (13.1 km/s)
Inclination of axis at poles: 3° 12′
Inclination of orbit from ecliptic: 1° 3′
Main atmospheric components (C = clouds): H/He, C
Average surface temperature day/night: –238°F (–150°C)
Gravity: 2.69 (Earth = 1)
Mass: 317.9 (1.899 x 1,027 kg) (Earth = 1)
Volume: 1.323 (Earth = 1)
Magnetic field: powerful
Known satellites: 63

Jupiter's Four Main Satellites

•EUROPA

Diameter: 1,942 miles (3,126 km)
Average distance from Jupiter: 416,900 miles (670,900 km)
Period of rotation (days): 3.551

• IO
Diameter: 2,257 miles (3,632 km)
Average distance from Jupiter: 261,970 miles (421,600 km)
Period of rotation (days): 1.769
• CALLISTO
Diameter: 3,026 miles (4,870 km)
Average distance from Jupiter: 1,139,000 miles (1,883,000 km)
Period of rotation (days): 16.689
• GANYMÈDE
Diameter: 3,278 miles (5,276 km)
Average distance from Jupiter: 665,000 miles (1,070,000 km)
Period of rotation (days): 7.155
• Other satellites of Jupiter : Adastrea, Amalthea, Leda, Himalia, Lysithea, Elara, Ananke, Carme, Pasiphae, Sinope, 1979-J2, 1979-J3, etc.

Saturn and its Satellites

Saturn

Diameter: 74,600 miles (120,000 km)
Average distance from Sun: 887,000,000 miles (1,427,000,000 km)
Minimal distance from Earth: 773,000,000 miles (1,244,000,000 km)
Period of revolution (years): 29.46
Period of rotation (days): 0.43 (10 h 15′)
Average orbital speed: 6 miles/second (9.7 km/s)
Inclination of axis at poles: 26° 44′
Inclination of orbit from ecliptic: 2° 5′
Main atmospheric components (C = clouds): H/He, C
Average surface temperature day/night: –292°F (–180°C)
Gravity: 1.16 (Earth = 1)
Mass: 95.2 (5.684 x 1,024 kg) (Earth = 1)
Volume: 752 (Earth = 1)
Magnetic field: powerful
Known satellites: 30

Saturn's Five Main Satellites

• TITAN
Diameter: 3,100 miles (5,000 km)
Average distance from Saturn: 745,000 miles (1,200,000 km)
Period of rotation (days): 15.95
• RHEA
Diameter: 950 miles (1,530 km)
Average distance from Saturn: 327,300 miles (526,800 km)
Period of rotation (days): 4.52
• IAPETUS
Diameter: 895 miles (1,440 km)
Average distance from Saturn: 261,970 miles (421,600 km)
Period of rotation (days): 1.769
• DIONE
Diameter: 695 miles (1,120 km)
Average distance from Saturn: 1,170,000 miles (1,883,000 km)
Period of rotation (days): 16.689
• TETHYS
Diameter: 650 miles (1,050 km)
Average distance from Saturn: 665,000 miles (1,070,000 km)
Period of rotation (days): 7.155
• Other satellites of Saturn : Mimas, Enceladus, Phoebe, Hyperion, 1980-S1, 1980-S3, 1980-S6, 1980-S13, 1980-S25, 1980-S26, 1980-S27, 1980-S28, etc.

Uranus

Diameter: 31,763 miles (51,118 km)
Average distance from Sun: 1,783,090,000 miles (2,869,600,000 km)
Minimal distance from Earth: 1,594,000,000 miles (2,565,000,000 km)
Period of revolution (years): 84.01
Period of rotation (days): 0.6 (17 h 24′)
Average orbital speed: 4.2 miles/second (6.8 km/s)
Inclination of axis at poles: 97° 86′
Inclination of orbit from ecliptic: 0° 8′
Main atmospheric components (C = clouds): H/He/CH4, C
Average surface temperature day/night: –345°F (–210°C)
Gravity: 1.17 (Earth = 1)
Mass: 14.6 (Earth = 1)
Volume: 67 (Earth = 1)
Magnetic field: powerful
Known satellites: 27

Uranus' Two Main Satellites

• TITANIA
Diameter: 685 miles (1,100 km)
Average distance from Uranus: 300,400 miles (483,400 km)
Period of rotation (days): 8.7
• OBERON
Diameter: 620 miles (1,000 km)
Average distance from Uranus: 351,900 miles (566,300 km)
Period of rotation (days): 13.46
• Other satellites of Uranus : Miranda, Ariel, Umbriel, etc.

Neptune

Diameter: equator 31,403 miles (50,538 km), poles 30,820 miles (49,600 km)
Average distance from Sun: 2,794,100,000 miles (4,496,600,000 km)
Average distance from Earth: 2,652,600,000 miles (4,269,000,000 km)
Period of revolution (years): 164.8
Period of rotation (days): 0.7 (16 h 03′)
Average orbital speed: 3.4 miles/second (5.4 km/s)
Inclination of axis at poles: 29° 56′
Inclination of orbit from ecliptic: 1° 8′
Main atmospheric components (C = clouds): H/He/CH4/methane, C
Average surface temperature day/night: –365°F (–220°C)
Gravity: 1.22 (Earth = 1)
Mass: 17.2 (1.028 x 1,023 kg) (Earth = 1)
Volume: 54 (Earth = 1)
Known satellites: 2 (Triton and Nereid)

Pluto

Diameter: 1,490 miles (2,400 km)
Average distance from Sun: 3,670,000,000 miles (5,900,000,000 km)
Minimal distance from Earth: 2,605,000,000 miles (4,192,000,000 km)
Period of revolution (years): 247.7
Period of rotation (days): 6.03
Average orbital speed: 2.9 miles/second (4.7 km/s)
Inclination of axis at poles: 50°
Inclination of orbit from ecliptic: 17° 2′
Main atmospheric components (C = clouds): CH_4, C
Average surface temperature day/night: –382°F (–230°C)
Gravity: 0.20 (Earth = 1)
Mass: 0.00025 (6.6 x 1,023 kg) (Earth = 1)
Volume: 0.01 (Earth = 1)
Known satellites: 1 (Charon)

Is Pluto a planet? Many astronomers feel that if Pluto were discovered today, it would not be called a planet. We now know that the region beyond Neptune is full of icy objects orbiting the sun, some of them quite large. Sedna (2003VB12) was discovered in November 2003. Sedna is estimated to be about 1,100 miles (1,770 km) in diameter, and at present it is the coldest, most distant object known to be orbiting the sun.

Comets and Asteroids

Diameter: hundreds of feet to miles
Average distance from Sun: variable
Average distance from Earth: variable
Inclination of axis at poles: no poles
Inclination of orbit from ecliptic: variable
Main atmospheric components (C = clouds): variable
Gravity: variable
Mass: variable
Volume: variable

Milky Way and Galaxies

Diameter: 100,000 light-years*
Thickness: 16,000 light-years
Distance from galactic center to Sun: 32,000 light-years
Period of revolution: (Earth years): 225 million

* 1 light-year = the distance a photon (a light particle) travels in 1 year at a speed of 186,000 miles/second (300,000 km/s).

Index